EYE FOR AN EYE

EYE FOR AN EYE

Decoding Global Special Operations and Irregular Warfare: A Vision for India

PREM MAHADEVAN

wisdom tree

First published 2014

ISBN 978-81-8328-366-3

Published by
Wisdom Tree
4779/23, Ansari Road
Darya Ganj, New Delhi-110 002
Ph.: 23247966/67/68
wisdomtreebooks@gmail.com

Printed in India

Dedicated to DK211176, with high respect

Ob Säbel, ob Kanon', ob Kleingewehr uns dräut,
Auf, Ansbach-Dragoner, Auf, Ansbach-Bayreuth!
Drum, Kinder, seid lustig und allesamt bereit,
Auf, Ansbach-Dragoner, Auf, Ansbach-Bayreuth!

—*Der Hohenfriedberger Marsch*

Translation:
Whether sabers, whether cannon, whether muskets threaten us,
Up, Ansbach-Dragoons, Up, Ansbach-Bayreuth!
So, boys, be jolly and all ready to go,
Up, Ansbach-Dragoons, Up, Ansbach-Bayreuth!

—*The Hohenfriedberger March*

Contents

Acknowledgements *ix*

Introduction *xi*

Part I: Global Overview

1. Tasking 3
2. Intelligence 25
3. Logistics 45
4. Training 67

Part II: Threat and Response

5. Pakistan's Proxy War 87
6. The Bear's Claws in Afghanistan 109
7. Israeli SOF and Counterterrorism 127
8. Synergising PSYOPS with Special Operations 147

Bibliography *169*

Index *181*

Acknowledgements

This book could not have been written but for the strong support of the Center for Security Studies (CSS) at ETH Zürich. I wish to particularly thank Prof Andreas Wenger, Dr Oliver Thränert, Dr Martin Zapfe and Dr Daniel Möckli for their encouragement and interest in my research. It goes to the credit of ETH Zürich that a study, which has aroused the interest of government policymakers half a world away, was authored within the intellectual landscape of Swiss-German academia.

The Indian Army's Centre for Land Warfare Studies (CLAWS) must be complimented for having commissioned this project. I am delighted to have worked full-time on an issue that has preoccupied my attention for well over a decade, ever since Pakistan-sponsored terrorism began to be visited with monotonous regularity upon an unsuspecting Indian population after the Kargil aggression of 1999. CLAWS, which represents the cutting edge of Indian military intellect, has long conducted exemplary research on special operations, and it was indeed a challenge to identify an angle that has not already been explored before. The work done by Lt General (Retd) PC Katoch and Saikat Datta on the evolution and employment of Indian Special

Forces is a particularly useful reference point for researchers focused on studying the Indian geo-strategic threat environment.

Having no wish to replicate the excellent efforts of these two gentlemen, whose valuable comments and criticisms of my own research I have endeavoured to incorporate wherever applicable, I have confined my enquiry to the issue of combating cross-border terrorism. Thus, this book does not explore broader issues regarding the value addition of Indian special operations in wartime, their potential efficacy along the frontier with China, or their employment beyond Pakistan. Rather, it seeks to examine what the nature and conduct of special operations should be, in response to Pakistan's long-drawn and increasingly audacious proxy war. I believe that this topic in itself is worthy of focused academic attention, because ignoring it only seems to encourage greater impunity from the Pakistani military, intelligence and *jihadist* community.

Major General (Retd) Dhruv Katoch and the publications and management staff at CLAWS have been extremely supportive in guiding my work—my deepest thanks go to them. Brigadier (Retd) Gurmeet Kanwal also deserves special mention for having asked me to undertake this project. I hope that our Armed Forces and strategic community might find it useful reading. Lastly, I would like to thank the handful of officers within the Indian security community, who over many years, have obliged me by agreeing to informally discuss the question of retaliating to Pakistan's proxy war. As always, it is possible that their views may or may not converge with mine, but I firmly believe that the essence of scholarship is to allow for a multiplicity of opinions and voices. I look forward to continued engagement with informed colleagues on this subject, whether they agree with my findings or not.

Introduction

This book examines the contribution that special operations can make to combating cross-border terrorism. It suggests that such operations yield strategically decisive results, only if they are combined with a counterterrorist policy that is proactive at the diplomatic level. Besides gathering intelligence and degrading terrorist resources, special operations can undermine terrorist confidence. However, to do this, they need to be synergised with psychological operations.

Often, the failure of special operations to deliver enduring results against irregular adversaries, despite impressive short-term successes, is the result of an inability to humiliate the enemy both politically and ideologically. Doing this requires *identifying the enemy's reversible strengths, and innovatively striking at them.* When the enemy's own assets can be used against him, by hitting deep in his hinterland where he is underprepared for defensive action, the shock effect is tremendous.

Getting behind his front line requires locating a weakness in his defences. Since any reasonably competent enemy would not initiate hostilities without first minimising his vulnerability to retaliation, hitting him when and where he least expects, is a matter of innovation. New capabilities for deep penetration strikes have to be built from scratch, or improvised from existing resources, such that a theoretical vulnerability in the enemy's defences becomes an actual, exploitable one.

Outline of the Argument

The book is divided into two parts. Part I is a historical overview of the role played by special operations in countering irregular warfare. It studies four key variables that have a bearing on mission success rates: Tasking, Intelligence, Logistics and Training (which can be abbreviated as TILT).

Part II is an analysis of the Pakistani proxy war against India and options for countering it. It explains why the current Indian policy of relying primarily on diplomacy has not worked, and how Pakistani covert strategy has been shaped by the Soviet-Afghan War and the Arab-Israeli conflict. It concludes by advocating a larger but still calibrated role for special operations within Indian counterterrorism.

Both sections deal with the same core questions: How have other countries combated cross-border terrorism, and what lessons can India learn from their experience? The second section also looks at what Pakistan has already learnt from its own experience of promoting such terrorism. The conclusion reached in both parts of the book is: To defeat terrorists you have to copy their strategy, if not their tactics, and prove even better at pursuing this strategy than they are.[1]

This does not mean that India must necessarily respond to a terrorist strike like the one in Mumbai on 26 November 2008 by setting off bombs in Lahore, or shooting Western businessmen in Karachi. It does not require that New Delhi should contest the Pakistani monopoly over murdering innocent civilians. Rather, it means that should India choose to retaliate with special operations,

[1] Kutger, Joseph P, 'Irregular Warfare in Transition', *Military Affairs*, Vol. 24, No. 3 (1960), p. 116.

these must primarily aim to sharpen Pakistan's existing civil-military divide, by highlighting the army's inability to guard the country's borders. So far, the Pakistan Army has got away with practicing selective vigilance: It allows terrorists to cross over into India, but threatens nuclear war if Indian troops were to enter Pakistan, even in hot pursuit.

Use Special Operations to Discredit Pakistan Army

A single special operation that exposes the Pakistan Army for what it is—a mafia that extracts protection money from civilian governments, but then fails to provide any real security from either external or internal transgression—would work admirably in India's interest. It would strengthen democratic forces in Pakistan, weaken obstructionist actors in the national security establishment, and strike fear into *jihadist* planners who currently anticipate a long working life. Such an operation, being a self-contained act with precisely calibrated levels of force, would stop well below crossing Pakistan's nuclear redlines. Importantly, it would show that India too has redlines.

Underlying the argument of this book is a simple philosophy: To break the enemy's spirit, you need to fight him on his own terms, and still win. This was the formula that the British used to conquer India—proving to native rulers that they could fight and win on political ground where the rulers had thought they held all the advantages.[2] To break the will of the Pakistani security establishment to use terrorism as an instrument of policy, it is necessary to prove that terrorism will not keep Pakistan's current borders safe from the Indian military. India must strike deep within Pakistani territory, having manoeuvered around their nuclear blackmail in a measured response.

[2] Darwin, John, *After Tamerlane: The Rise and Fall of Global Empires, 1400-2000* (London: Penguin, 2007), p. 264.

A Note on Terminology

Before proceeding further, readers need to be familiarised with four key terms, around which the argument of this book is based.

Special operation: A time-sensitive mission, undertaken by selected personnel operating as a team in an area dominated by the enemy, for the purpose of neutralising a critical target or retrieving a political asset. Unlike an intelligence operation, the personnel do not work alone and over extended periods of time, but have a strict deadline to meet before the utility of their mission fades. At a minimum, their task is to gather intelligence by observation, prisoner-snatching, or document seizure. It could, however, also extend to neutralising an individual or installation on which detailed intelligence has already been collected by other means, both human and technical. Beyond these two categories of tasks—intelligence acquisition and direct action—special operations should be clearly distinguished from the personnel who carry them out: Special Operations Forces (SOF). Such personnel (often abbreviated to Special Forces) are capable of undertaking a broad range of activities, including those which can be handled by regular infantry or airborne/airmobile units. Their use in such a capacity does not qualify their combat actions as 'special operations'. Rather, such actions merely represent the use of SOF in conventional roles. A special operation is 'special' because it poses challenges which cannot be overcome by regular units, no matter how well-trained or well-led they are. It requires specialised skills, logistical arrangements, and command structures.[3]

Irregular Warfare: A category of armed conflict, wherein guerillas or terrorists based in one state, repeatedly plan and execute attacks on the territory of another state, thus creating a warlike situation.

[3] Moore, David M; Allen, David and Antill, Peter D, 'Strategy development for special operations force logistics', *Defence Studies*, Vol. 3, No. 1 (2003), pp. 67-69.

Even if they do not receive active support from the host government, they benefit from its tacit support. This is because in an age where national sovereignty is absolute, no state can claim inability to control persistent cross-border transgressions from its territory, without acknowledging that it has no domestic authority. Irregular warfare is, thus, an inter-state war that is not recognised as such by international law, but which can feature very high levels of violence approximating that of a regular military confrontation.[4] It is characterised by an asymmetry of methods, but not necessarily of willpower. The state assisting guerrillas/terrorists publicly denies its involvement in their activities, but privately leverages these for its own benefit. The victim state seeks to expose the aggressor, but without surrendering its own right to exclusively manage its internal affairs. In such warfare, the advantage lies not with the country that is more committed to its goal, but which manages to use its non-military influence to shape international perceptions of the other. The victim state faces a double challenge here: Not only does it have to prove external interference in its domestic affairs (a near-impossible task, given the nature of covert operations), but it also has to explain indigenous violence that has local roots, and keep the discourse focused on cross-border violence.

Reversible Strengths: Those assets of an enemy which can be abruptly turned to his disadvantage by applying an unorthodox method of attack for which he has not yet developed a response. For example, a rear-area assault will do considerable damage in a country that enjoys strategic depth, because vigilance levels drop as one moves further from combat zones. Intricate command and control systems, which help in coordinating large-unit offensives over long distances, stifle junior leadership during a localised and fast-

---

[4] Cassidy, Robert M, 'Counterinsurgency and Military Culture: State Regulars versus Non-State Irregulars', *Baltic Security and Defence Review*, Vol. 10 (2008), p. 53.

moving crisis. Standard operating procedures suited for containing one kind of threat can worsen another. The game-changing effect of reversible strengths has been visible whenever a suicidal gunman runs amuck in an American school, or a *fidayeen* terrorist attack hits a civilian locality. Special Weapons and Tactics (SWAT) teams have long been trained to respond to a shooting incident with a slow and deliberate buildup of force, in the hope that this would intimidate the attacker(s) to surrender quietly rather than be killed.[5] Such a pressure tactic works well with rational criminals and terrorists who want to continue living, but is counterproductive when applied against deranged psychopaths or suicidal terrorists. It only allows them more time to kill more people. What was an asset in one situation—slow reaction—becomes a liability in another.

Exploitable weaknesses: Those vulnerabilities in an enemy's defences which he is unaware of, and against which he, therefore, has no safeguards. Such vulnerabilities are usually fleeting; for example, an informer who betrays one of his friends, a fugitive from the law, in exchange for a reward. In this case, the moment of weakness lasts only as long as the fugitive remains in contact with the informer and the informer in turn is in contact with his handlers; once either contact is broken, the fugitive is no longer vulnerable. On rare occasions, vulnerability can also be structural. New technologies can open up a state to sustained penetration, either physically (as with stealth aircraft) or electronically (as with cyber espionage). In these situations, the enemy becomes totally defenceless: He cannot protect himself from a deep-penetration strike any more than a shallow-penetration raid. Exploiting a weakness, however, whether it is fleeting or structural, requires tenacity and imaginativeness. The enemy's existing defences have to be minutely scrutinised for cracks through

[5] Klein, George C, 'Thinking SWAT', *Journal of Police Crisis Negotiations*, Vol. 3, No. 1 (2003), pp. 75-77.

which an attack can be mounted—an effort that requires a long-term commitment of both, intelligence resources and planning time. The payoff might only be a one-off because each success in breaking through an enemy's defences hinges on the enemy not recognising his own vulnerability. However, it is exploitable weaknesses that allow special operations to have a chance of succeeding in the first place.[6]

Role of Special Operations in Combating Irregular Adversaries

There are two opposed schools of thought on special operations: That they are critical to success in irregular warfare, and that they contribute little to it. The first school argues that because special operations feature small-unit tactics and unorthodox attack techniques, and rely on intelligence for their success, they are a key indicator of an effective counter-insurgency or counterterrorism campaign.[7] The second school believes that because irregular adversaries do not present fixed targets which can be neutralised by direct action, special operations cannot do them much harm. This argument suggests that unlike conventional forces, which can be hit at logistical chokepoints, guerrillas have few physical assets whose destruction can cripple their capacity for action.[8]

The disagreement arises mainly from differing perspectives. Special operations can indeed be highly effective against guerrillas, but only when carried out within the guerrillas' own strongholds. Areas where irregular fighters have set up recruitment, training and

---

[6]Gray, Colin S, 'Handfuls of Heroes on Desperate Ventures: When do Special Operations Succeed?', *Parameters* (Spring 1999), http://strategicstudiesinstitute.army.mil/pubs/parameters/Articles/99spring/gray.htm, 16 September 2013.

[7]Brailey, Malcolm, *The Transformation of Special Operations Forces in Contemporary Conflict: Strategy, Missions, Organisation and Tactics* (Duntroon, Australia: Land Warfare Studies Centre, 2005), p. 26.

[8]Thomas, David, 'The Importance of Commando Operations in Modern Warfare 1939-82', *Journal of Contemporary History*, Vol. 18, No. 4 (1983), p. 703.

logistics infrastructure, however rudimentary these are, become target-rich environments for special operations forces. If, on the other hand, SOF are employed within areas where government troops are already dominant, then they contribute little to the overall campaign. Such usage of highly trained assets would be a gross mistake and would not qualify as special operations per se. From what can be observed by studying these operations across time and space, three generalisations can be made:

EACH SPECIAL OPERATION IS UNIQUE, AND SHOULD REMAIN SO

Perhaps the biggest mistake that policymakers make is in thinking that successful special operations can be replicated. Each operation is different, whether it ends in success or failure. All that mission planners can do is have a checklist of parameters to gauge their preparedness to carry it out.[9] They cannot guarantee the outcome, much less guarantee further success in subsequent operations of the same type. This is particularly true of special operations carried out in peacetime, which are subjected to much greater political constraints than wartime operations. Positive results are not assured.

---

[9]The best such checklist has been drawn up by William McRaven, the US Navy officer who masterminded the 2011 SEAL raid on Osama bin Laden's hideout at Abbottabad. McRaven argued that for a direct action mission to succeed, the attacking force has to swiftly gain 'relative superiority' over the defending force. He defined relative superiority as a tactical condition wherein attackers can cripple the defenders' ability to interfere with the objective of the mission. Relative superiority is attained at the pivotal moment of an engagement, which is usually the moment of greatest risk, and once gained, needs to be retained by a combination of boldness and perseverance on the part of the attackers. If lost, it is difficult to regain and the probability of mission failure increases dramatically. To acquire relative superiority, the attackers must have a simple attack plan, rehearse it extensively amid tight operational security, adopt innovative tactics and technologies to surprise the enemy, execute the mission rapidly so as to deny the enemy an opportunity to interfere with it, and retain a sense of purpose even when unforeseen eventualities complicate the implementation. See McRaven, William H, *Spec Ops: Case Studies in Special Operations Warfare – Theory and Practice* (Novato, CA: Presidio, 1996), pp. 6-23.

Even so, there is an initial tendency to view special operations as a 'silver bullet' in counter-insurgency/counterterrorism. When real-world considerations of intelligence and logistics intervene, much of this enthusiasm dries up. It is felt then that if special operations require so much support from earmarked resources, which have to be pulled away from other duties, then it is better to place minimal reliance on such operations. This view is obstructive, but not unhealthy. History suggests that overuse of special operations is a worse mistake than underuse.

A ready example is that of the US intervention in Somalia in 1992-93. Once the Americans defined their mission as the neutralisation of Somali warlords, they began using special operations to conduct 'snatch' operations in the capital city, Mogadishu. After six such missions, the Somalis figured out how the Americans operated. US SOF tried to vary their tactics, particularly regarding modes of ingress and egress, in an effort to confuse Somali spies who were observing their activities. However, these innovations were confined within an operational template that became predictable over time. Such predictability allowed the Somali warlord Mohammad Farrah Aideed and his advisors to develop a plan to ambush the Americans, the next time they launched a raid.

The result was 'Black Hawk Down'.[10] Having realised that the US crucially depended on air power for its special operations, Aideed's men developed a simple yet effective scheme: They shot down two US helicopters, forcing the Americans to fight their way through the city's streets in order to rescue the aircrew. Unprepared for such a contingency, US soldiers were attacked from every direction by swarms of Somali gunmen. Although the death toll was extremely high for the Somalis, they were still able to fight the Americans,

[10] Bowden, Mark, *Black Hawk Down* (London: Corgi, 2000), p. 42.

man against man, and not man against aircraft. The ability of a group of rag-tag fighters to develop such a battle plan stemmed from an innate American tendency to overuse special operations, to the point where their weaknesses became evident and susceptible to exploitation.

The Somalia example highlights an additional point: Besides numerical inferiority, special operations forces are disadvantaged by lack of local knowledge. The original 'commandos' of the Boer War, who fought the British colonial army in South Africa, had operated in their own backyard. They knew the lie of the land much better than the more numerous and better-armed British, and were better fighters on an individual basis. Taken together, these assets of terrain knowledge and combat skill allowed them to attack the British suddenly, using secret routes, and then melt away from the engagement area faster than the British could regain their wits and organise a pursuit.[11]

Today's SOF commandos do not have comparable levels of terrain knowledge, since they are operating on unfriendly territory. This means that only at an initial stage will their actions have a shock effect. Once the enemy has regained his composure, he will be able to use his better ground-level information to organise a pursuit of the raiding forces. Therefore, special operations in enemy territory should not become a habit, to the extent that countermeasures can be devised to tackle them.

STRANGULATION IS THE MAIN OBJECTIVE OF BOTH SIDES IN IRREGULAR WARFARE

Both governments and guerrillas are dependent on support networks, moral and material. When these networks are shut down on one

[11] Dobbie, Elliot VK, 'The Word "Commando"', *American Speech*, Vol. 19, No. 2 (1944), pp. 81-89.

side, while remaining open for the other, the strategic balance shifts decisively. This is because in irregular warfare *attrition* and *annihilation*—the two main strategies of conventional warfare—play only a limited role. Instead, it is the slow but steady *isolation* of the enemy's power centres and the denial of extraneous support to his means of resistance that eventually leads to his destruction or capitulation.

Weakening an irregular adversary requires that a government attack the connections between his sympathisers and the operational cadres that actually do the fighting. The British found a way of doing this in counter-insurgency: Concentration camps. During the Boer War, tens of thousands of Boer women and children were seized from their homes and interned en masse in squalid open-air prisons, where starvation and disease killed many. Their removal from the operational area deprived Boer commandos of food, information and shelter. Moreover, knowledge that their families were dying because of British ill-treatment eroded the commandos' morale.[12] Similar methods were used to starve communist insurgents in Malaya during the 1950s—large numbers of ethnic Chinese were uprooted from their homes and forcibly moved into 'resettlement' camps (although admittedly, living conditions here were much better than had been in the case of the Boer concentration camps).

When it comes to counterterrorism, the main aim of governments is not to deny terrorists food or shelter, but legitimacy. Likewise, the terrorists' aim is to undermine the legitimacy of current government policies. Towards this, terrorists use special operations as a psychological warfare tool, by carrying out high profile

---

[12] Downes, Alexander B, 'Draining the Sea by Filling the Graves: Investigating the Effectiveness of Indiscriminate Violence as a Counterinsurgency Strategy', *Civil Wars*, Vol. 9, No. 4 (2007), pp. 427-437.

attacks which generate a lot of media coverage, and then subtly steering this coverage towards criticism of the government.[13] One of the first groups to synergise special operations with psychological operations was Hizballah, which back in the 1980s, introduced the concept of mass casualty suicide bombing. Realising that such attacks prompted introspection within Western governments, Hizballah began to manipulate Western discourse on terrorism. It released pre-recorded videos of suicide bombers, in which they explained their motives for attacking Westerners by citing personal grievances that collaterally also pushed forward Hizballah's own agenda.[14] By drip-feeding its narrative into the media spotlight, Hizballah weakened the legitimacy of its enemies' policies.

Al Qaeda copied this strategy. From the late 1990s onwards, it transferred several cadres from its propaganda departments to its special operations unit.[15] The result was a drift towards carrying out spectacular attacks which would generate publicity for the group. Al Qaeda also reached out to its sympathisers via cyberspace, knowing that publicity provided an opportunity for putting its message out. Even as Western intellectuals asked, 'Why do they hate us?' Al Qaeda ensured that the answers focused more on the historical misdoings of the West rather than its own more recent atrocities. It, thus, employed special operations to gain international attention, and

---

[13] Terrorists have understood two basic principles of psychological operations: 1) they must be offensive, i.e., they must seize the initiative in shaping the narration of broader trends and the preparation of historical accounts, and 2) the messages that they transmit must reach the targeted individual through his own circle of intimate friends and associates. This is where the international media, which is supposedly neutral and unbiased, comes in. Pahlavi, Pierre, 'Political Warfare is a Double-edged Sword: The Rise and Fall of the French Counterinsurgency in Algeria', *Canadian Military Journal* (Winter 2007-2008), p. 59.

[14] Bergman, Ronen, *The Secret War with Iran* (London: Oneworld, 2008), p. 221.

[15] Gunaratna, Rohan and Oreg, Aviv, 'Al Qaeda's Organizational Structure and its Evolution', *Studies in Conflict & Terrorism*, Vol. 33, No. 12 (2010), p. 1064.

then used propaganda to shape the narrative that was formulated as a result of that attention.[16]

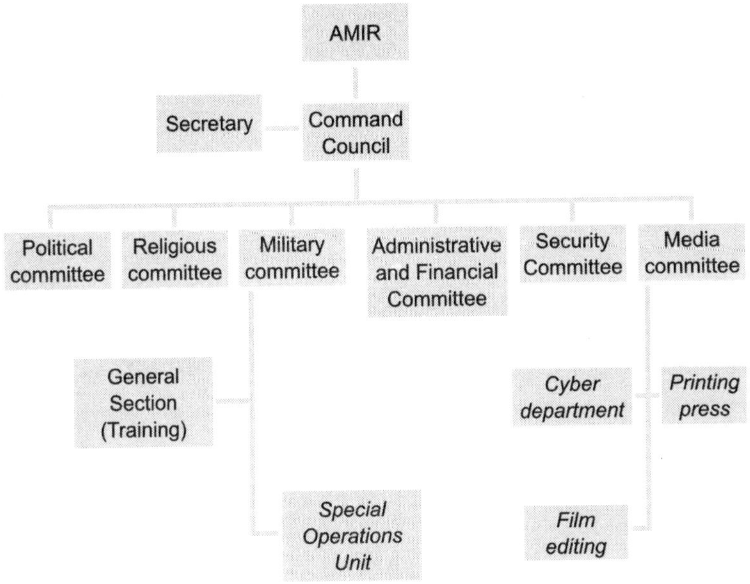

Diagram: Al Qaeda's organisational structure. The highlighted boxes show the cross-departmental posting of cadres between propaganda duties and planning for special operations.

On the counterterrorist side, the Israeli Mossad was the first to figure out how to play this game in reverse, by using special operations to discredit terrorists among their own sympathisers. Worried by the spate of airline hijackings that targeted Israel during the early 1970s, Mossad analysts concluded that special operations forces needed to conduct a high profile rescue mission against seemingly hopeless odds. The shock effect of such a mission, magnified by coverage in the international media, would discredit hijacking as a politico-

[16]Osama bin Laden said that in some cases, media management might be as much as 90 per cent of the preparation required to ensure that an Al Qaeda attack is successful. See Bergen, Peter and Cruickshank, Paul, 'Revisiting the Early Al Qaeda: An Updated Account of its Formative Years', *Studies in Conflict & Terrorism*, Vol. 35, No. 1 (2012), p. 24.

psychological warfare instrument. Thus, they were quick to perceive an opportunity when Palestinian terrorists hijacked an Air France passenger aircraft to Entebbe in 1976. The Mossad chief personally lobbied Israel's prime minister to authorise a rescue mission, knowing that its success would humiliate the terrorists among the larger Palestinian community. The result was Operation Thunderbolt, remembered even today for its daring.[17]

SPECIAL OPERATIONS ARE MORE THAN JUST SMALL-UNIT TACTICS

There is an intriguing possibility that advocates of special operations are not motivated by a conviction that these operations are in themselves useful for fighting irregular adversaries. On the contrary, their calculations might hinge on a view that special operations, and the personnel who participate in them, are so peripheral to 'mainstream' soldiering that they should only be confined to irregular warfare. Readers of this book would know that among career military officers, there exists a vague disdain for special operations. Often, personnel who get into SOF units are not the brightest, but those who have difficulty adjusting to life within a regular military unit, and seek to escape this by a transfer. In this situation, it is not surprising that special operations tend to be seen as dominated by misfits who are unwanted by their parent units.[18] Although there are exceptions to this pattern, and indeed, SOF training is geared towards poaching capable soldiers from regular units (causing resentment within the larger military), an institutional mentality that considers SOF as suitable only for low-priority tasks is still not hard to find. Combined with the fact that most armies regard counter-insurgency and counterterrorism as a secondary mission, not comparable in importance to conventional war-fighting, this suggests

---

17 Thomas, Gordon, *Gideon's Spies: Mossad's Secret Warriors* (London: Pan, 1999), p. 149.

18 Connelly, Mark and Willcox, David R, 'Are You Tough Enough? The image of the special forces in British popular culture, 1939-2004', *Historical Journal of Film, Radio and Television*, Vol. 25, No 1 (2005), pp. 7-8.

that bureaucratic enthusiasm for SOF usage in irregular wars has a lot to do with sparing regular military units from this task.[19]

The perceived value-added of special operations forces, according to their advocates, partly stems from their organisation: SOF typically operate in small-sized units. They are, thus, more adaptable to changing local conditions than infantry battalions; a quality that is useful when fighting irregular adversaries who do not have a fixed order of battle. However, small-unit tactics are not the preserve of SOF alone. Even regular infantry units can be retrained for such tactics, if provided with time and institutional backing. In the 1970s, Rhodesia (now Zimbabwe) reoriented its entire military to wage counter-insurgency once it realised that conventional war posed a lesser risk than domestic insurrection.[20] As a result, tactical speed and stealth increased: Rhodesian soldiers could neutralise insurgents even before the latter became aware of their presence in an operational area. By moving in small teams, army units were able to avoid alerting rebel spies, who were only looking out for large troop deployments. The same was true for British counter-insurgency in Malaya.[21]

Special operations play a bigger role in irregular warfare than do small-unit tactics: The latter are only employed within one's own territory, whereas special operations can also be carried out across international borders. In this, they serve as a vital adjunct to intelligence collection.[22] They gather information directly through physical observation of insurgent bases and training camps. They also

[19] Cassidy, Robert M, *Counterinsurgency and the Global War on Terror: Military Culture and Irregular War* (Stanford, CA: Stanford Security Studies, 2008), p. 50.

[20] Hoffman, Bruce; Taw, Jennifer M and Arnold, David, *Lessons for Contemporary Counterinsurgencies: The Rhodesian Experience* (Santa Monica, CA: RAND Corporation, 1991), pp. 2-4.

[21] Ball, Rhys, 'The strategic utility of New Zealand Special Forces', *Small Wars & Insurgencies*, Vol. 22, No. 1 (2011), p. 122.

[22] Stringer, Kevin, 'Historical Lessons from the British and Germans: The case for integrated indigenous forces as a logial force multiplier', *Canadian Military Journal* (2007), p. 61.

indirectly contribute by providing prisoners with a high interrogation value, and by capturing rebel documents. Since special operations occur in the enemy's heartland, where he is complacent and less mindful of personnel, communications and document security, they offer higher intelligence returns than engagements fought in more contested areas. However, to be directed at the right targets, they also require more intelligence support than do conventional operations.[23]

Over a period of time, special operations can assist intelligence monitoring of irregular threats in a way that no other military asset can. They relieve national intelligence agencies of the need to track terrorist and insurgent groups closely, by taking over part of that responsibility for themselves. A self-perpetuating loop is created between 'actionable' intelligence and special operations, wherein one reinforces the other. New techniques of agent-running, optimised for local conditions, can be introduced at the field level without requiring systemic intelligence reforms. Thus, national intelligence agencies can focus on monitoring traditional (i.e., conventional) threats at the strategic level, while special operations forces handle irregular ones at the tactical level.[24]

Relevance for India

Following the 26/11 Mumbai attacks, the Indian Army's readiness to launch a limited offensive against Pakistan was severely tested.[25] Failure to take punitive action, despite clear evidence of state involvement in the terrorist assault, was a serious policy mistake. To be fair, the Indian political leadership was coerced by the US and

---

[23] Cline, Lawrence E, *Intelligence, Special Operations and Counterterrorism: Improving the Links*, Paper delivered at the International Studies Association Annual Convention, March 2004, pp. 2-5.

[24] Charters, David A, 'Special Operations in Counter-Insurgency: The Farran case, Palestine 1947', *RUSI Journal*, Vol. 124, No. 2 (1979), p. 57.

[25] Raza, Maroof, 'Introduction', in Raza, Maroof ed., *Confronting Terrorism* (New Delhi: Penguin, 2009), pp. xxiv-xxv.

the UK, who went all-out to protect Pakistan from any reprisals. These countries, together with Australia, Canada and New Zealand, airbrushed and initially even denied outright any possibility that state agencies within Pakistan could have been complicit in the attacks, a line of argumentation that we now know to have been false.[26]

In the years since, the US and the UK have slowly retreated from their assurances of securing Pakistani cooperation in counterterrorism. They have basically abandoned India to cross-border *jihadist* whims, while professing goodwill and strategic assistance on the basis of 'shared democratic values'. It is necessary that our national security establishment recognise these assurances as insincere, and reduce dependency on the West for fighting terrorism.[27] India needs to find its own solutions, and special operations should be an integral part of any policy package for dealing with Pakistan.

RESPONDING TO NUCLEAR BLACKMAIL

Pakistan is fond of dropping hints that any Indian response to a state-sponsored act of terrorism, even if scrupulously limited to military targets, will invite 'massive retaliation'.[28] This includes special operations as well as surgical air strikes on terrorist camps. Although such nuclear blackmail lacks credibility if closely examined, it still succeeds in intimidating the West, which in turn pressures India to avoid carrying out even a token reprisal. Therefore, Indian retaliation must begin as a covert operation, which is not claimed by New Delhi. Over a period of time, the diplomatic, intelligence and logistical groundwork can be laid for a series of direct action missions, to be carried out at a time of Delhi's choosing. Instead of waiting to

---

[26] 'WikiLeaks reveals how foreign diplomats charted response to 26/11', *Indian Express*, http://www.indianexpress.com/news/wikileaks-reveals-how-foreign-diplomats-charted-response-to-2611/719127/, 9 September 2013.

[27] Pathak, DC, 'Understanding the "War on Terror"', *Eternal India* (2009), pp. 53-54.

[28] Kumar, A Vinod, 'A Year after 26/11: Soft Responses of a Reluctant State', *Strategic Analysis*, Vol. 33, No. 6 (2009), p. 804.

be hit, India must seize the initiative and commence target-specific preparations for retaliation, even before the next attack occurs.

LACK OF A SPECIAL OPERATIONS TRADITION

Unfortunately, India does not have a strong tradition of special operations to draw upon. In the Indian Army, such operations are still conflated with small-unit tactics. Doctrinally, the Army retains a conventional warfare orientation, and rightly so—it needs to prioritise fighting the Chinese and Pakistanis in regular combat, since interstate war remains very much a part of the regional threat spectrum.[29] However, this also means that India places a lower overall priority on fighting irregular warfare. The basic fighting unit in Indian counter-insurgency operations is a company-sized force, which can carry out raids, cordon and search operations, and area patrolling, all without changing its character. Its main task is not to hunt down insurgents but to keep arterial roads and other supply routes open, in the event that a sudden build-up is required on the borders with China or Pakistan.

A surfeit of manpower in such units leads to a lack of tactical initiative and generates tolerance for poor combat skills, especially in marksmanship. With so many soldiers pursuing a relatively small number of insurgents, the volume of ammunition expended in engagements tends to vastly exceed the level required, especially when contrasted with restrained and targeted shooting.[30] In this situation, a premium is placed on troops that have high stamina and above-

[29] History seems to suggest that counterinsurgency-centric armies have difficulties in adapting to conventional war-fighting, but the reverse does not apply: A competent army can indeed, adapt to counterinsurgency if required to. It was for this reason that the British army, which was basically a counterinsurgency-centric force, did quite poorly in conventional warfare in both World Wars, and only won because of the immense advantages the British Empire had in manpower, allies and resources. See Porch, Douglas, 'The dangerous myths and dubious promise of COIN', *Small Wars & Insurgencies*, Vol. 22, No. 2 (2011), p. 251.

[30] Rajagopalan, Rajesh, *Fighting Like a Guerrilla: The Indian Army and Counterinsurgency* (New Delhi: Routledge, 2008), pp. 98-99 and pp. 129-130.

average combat skills. These troops are grouped as 'Special Forces' and deemed suitable for missions that are euphemistically labelled 'special operations' but are actually small-unit actions, of the kind carried out by both regular British infantrymen and specially trained jungle commandos during the Malayan Emergency.

TESTING EMPIRICAL CONCLUSIONS

This book provides an overview of the role that special operations can play in combating Pakistan-sponsored terrorism, based on the use of such operations elsewhere. It tests the conclusions reached by Bruce Hoffman in his landmark 1985 study, *Commando Raids: 1946-1983*. The study empirically analysed 100 special operations to assess the potential contribution that commandos could make to counterterrorism. Hoffman concluded that special operations can play a positive role if they are integrated within a larger counterterrorist strategy. His conclusions were derived from patterns of mission success across two variables:

- Type of mission (domestic, cross-border or international) and
- Type of personnel employed (elite units or irregular fighters).[31]

His findings are summarised below:

Mission success rate in percentages (total sample size: 100 raids)

	Irregular Fighters	Elite Units	Overall
Domestic	91% (10/11)	100% (5/5)	94% (16/17)
Cross-border	56% (19/34)	94% (34/36)	76% (53/70)
International	80% (4/5)	50% (4/8)	61% (8/13)

The figures in brackets indicate the actual number of successful raids in each category, relative to the total number attempted.

[31] Hoffman, Bruce, *Commando Raids: 1946-1983* (Santa Monica, CA: RAND Corporation, 1985), http://www.rand.org/content/dam/rand/pubs/notes/2005/N2316.pdf, 16 September 2013.

Hoffman's study suffered from one shortcoming, in that it looked only at direct action missions. This book goes further by studying the overall contribution that special operations can make in combating irregular adversaries, extending up to the level of grand strategy. It suggests that special operations are not just useful in ground-level counterterrorism, but they are also essential to Indian national security in a holistic sense. Faced with both conventional and irregular warfare threats of mounting intensity, the Indian Army needs to develop a capability to strike deep within Pakistan, using unorthodox methods presently not considered as part of conventional war-planning. Regular military forces can deter Pakistan from carrying out another Kargil-type misadventure, but only targeted neutralisation of key terrorist planners and logisticians will put a stop to nuclear weapon-enabled terror. This is where special operations will have a crucial role to play in years ahead.

Part I

Global Overview

The mission of the Special Forces soldier was hard, dirty, and bloody. Far from the air-conditioned staff offices and comfortable bungalows of higher headquarters, the Special Forces advisors were crammed into dirt-walled, undergunned campsites carved out of steaming, disease-ridden swamp and jungle. There was little respite from constant danger and back-breaking labor in the dank, mist-shrouded hinterland of Vietnam. Sinewy muscles were toned and hardened by the long patrol marches over broken grassland, leg-wrenching-deep mud, and vine-choked jungle. The sweltering tropical heat baked away sweat and blistered eyelids almost shut. Foul water and malaria-infested rainforest sent burning fever and clammy chills through bodies doubled over by rucksacks of ammunition strapped under radios. Sores openly festered as lice and insects crawled across skin slashed by thorns and razor-sharp elephant grass. Deadwood and animal-trodden rice fields twisted ankles and caved in knees, while around-the-clock vigilance reddened eyes and fatigued the mind. Sudden death or crippling disfigurement threatened every measured Special Forces step.[1]

—Shelby L Stanton (US military historian)

---

[1] Stanton, Shelby L, *Green Berets at War: US Army Special Forces in Southeast Asia, 1956-1975* (New York: Ivy Books, 1985), pp. 93-94.

1

Tasking

This chapter argues that special operations are decisive when combined with an offensive strategy—one that is uniformly proactive at the military, political and diplomatic levels. Such a strategy needs to treat special operations not as a sideshow, but as an essential ingredient for victory. In irregular warfare, this means that in order to influence the cost-benefit calculus of insurgent/terrorist groups and their state patrons, special operations need to be integrated with security campaigns at the domestic level, and with diplomatic efforts at the international level.

The chapter argues that the current conception of special operations—as a propaganda tool whose main contribution is to boost morale—is misguided. Special operations can crucially determine the outcome of a conflict, provided they receive full support from national policymakers. However, to build such a whole-of-government consensus is very difficult with the result that special operations are often unable to generate strategic impact. Besides political indecision, there is a strong bureaucratic reason to avoid focusing on special operations; mission failure can destroy careers.

First, the chapter shall differentiate between two models of special operations: The British and the German. Both took shape during World War II (WWII), but the former has completely eclipsed the latter in years since. Yet, its utility is considerably lower from a military perspective. The chapter shall then distinguish between three types of role that special operations play against irregular warfare: Area denial, area dominance and area penetration. It shall suggest that area dominance is perhaps the role that yields the highest return on investment, due to a host of enabling factors. Finally, the chapter shall conclude with an analysis of how the US built the capacity to carry out one of the most spectacular area penetration missions of all time: the killing of Osama bin Laden.

Difference between British and German Special Operations in WWII

At the tactical level, all special operations are offensive, but at the strategic level, the British have traditionally used special operations as the offensive prong of a defensive posture. Right from the first commando raids on Nazi-occupied Europe in WWII, the British have treated special operations as a device to boost public and military morale. The military value of such operations is not considered as important as their politico-psychological effect.[1] In part, this view of special operations has arisen from an inferiority complex over the British Army's poor performance in land warfare.

Having been built as a colonial and a counter-insurgency force, the British Army has long been used to fighting opponents who are less well-armed and well-trained. Thus, when routed by the highly capable German Wehrmacht in 1940, it was confronted at home with

[1] Weale, Adrian, *Secret Warfare: Special Operations Forces from the Great Game to the SAS* (London: Coronet, 1997), p. 274.

a public disillusioned by the ineptitude of its leaders. In this context, the army backed a proposal by Prime Minister Winston Churchill, to raise a special force to carry out pin-prick raids on German forces. For the remainder of the war, units like the Commandos and Special Air Service (SAS) basically served as showpieces whose combat actions were primarily intended to raise morale and only secondarily to assist conventional operations.[2]

German Wehrmacht, on the other hand, was a dynamic and inventive force with a far superior intellectual leadership. Its success was based on a new doctrine of manoeuvre warfare called Blitzkrieg (lightning war). Part of this doctrine called for specially selected and equipped soldiers to be grouped into anti-sabotage teams and inserted behind enemy lines just prior to the onset of hostilities. At a pre-appointed time, they were to seize logistical chokepoints along the enemy's anticipated line of withdrawal, keeping them open for advancing German tanks and mechanised infantry columns to use as the Wehrmacht pushed faster and deeper into enemy territory than the enemy's own forces could withdraw, thereby cutting off his retreat and forcing him to surrender.

The German way of war called for special operations to target an enemy's reversible strengths while conventional forces struck at his exploitable weaknesses. By seizing critical infrastructure such as bridges in rear areas, German special mission teams were able, at a stroke, to turn the enemy's assets into liabilities which would assist their own campaign objective. This approach worked well because the Wehrmacht had limited resources but ambitious political objectives, and thus tended to use its capabilities in the most cost-effective way possible. The British, on the other hand, had the manpower resources of a global empire and multiple allies to draw on, and therefore, used

[2] Connelly and Willcox, 'Are You Tough Enough?', p. 6. It needs to be mentioned that the Commandos were raised in June 1940, primarily for operations in Europe, while the SAS was raised a year later, in July 1941, to conduct raids behind enemy lines in North Africa.

special operations as a gap-filler while building up their conventional forces for attritional campaigns.[3]

During the war, British special operations were more effective at gathering intelligence on the enemy, than at degrading his combat capability. German special operations, on the other hand, were highly effective at crippling the enemy militarily, and thus, played a greater role in influencing the outcome of theatre campaigns. Unfortunately, with Germany having lost the war, much of the established thinking on special operations has since been shaped by British assessments. The SAS in particular has been treated as a gold standard to be emulated, although it is actually a white elephant whose activities offer few lessons for India, by way of a solution to cross-border terrorism.[4]

Instead of a showpiece unit which it can ill-afford, India needs to devise a special operations concept that would fit into its routine security and diplomatic efforts.[5] It needs to simultaneously *deny*

---

[3] Thomas, 'The Importance of Commando Operations in Modern Warfare 1939-82', p. 700.

[4] Gupta, Shekhar, 'Blood, Sweat and Tears', in *The Punjab Story* (New Delhi: Roli, 2004), p. 99.

[5] Special operations forces could play non-kinetic roles in promoting Indian interests, but these roles are not relevant to task of retaliating to cross-border terrorist attacks. Saikat Datta, in a seminar discussion on this book on 17 December 2013, made the point that the Special Air Service helped re-establish British influence in the Middle East after the 1956 Suez Canal crisis. Having been politically humiliated during the crisis, the British used SAS training missions as a vehicle for building new ties with postcolonial Arab governments. This author is sceptical of using Indian SOF in a similar role, for two reasons. First, the UK is a has-been on the world stage, that after 1947 has used the SAS to make up for its weakening as an expeditionary military actor. The employment of the SAS in military diplomacy is a symptom of overall British feebleness, not strength. India would need to emulate this use of SOF only sparingly, in theatres such as Afghanistan and perhaps Myanmar. The second reason for scepticism is that major Western powers actively monitor and circumscribe Indian power projection efforts. We can assume that India's growing international profile, especially if fronted by serving or retired military personnel (whether training teams, combat units, or even private security contractors) would not be welcomed by Western powers such as the US and the UK. These countries are clear that their long-term policy is to act as offshore balancers in Asia and they presently wish to hedge against the emergence of any strong Asian state. Should India appear to be proactively pushing forward its national interests using military diplomacy, Western powers would likely step in to offer alternative services to the local political and military elites that New Delhi is courting, in order to curtail our influence.

Pakistani terrorists space to operate in its territory, *dominate* the remote localities where they hide, and *penetrate* their base areas in Pakistan itself. However, as this chapter shall show, not all these functions are equally effective. Historically, both area denial and area penetration have yielded limited benefit, and are mostly justified in terms of their low cost outlay (area denial) or psychological effect (area penetration). The most optimal use of special operations seems to be in area dominance.

Area Denial Missions

Strictly speaking, area denial is more a training mission than a special operation, since it is not a time-bound task. However, it has been included in this book because the training of anti-insurgent/anti-terrorist militias is often a crucial part of the mandate allocated to special operations forces. The skills required to be a good trainer are also quite specialised, thus requiring that area denial responsibilities should not be accorded less attention than more exciting or 'kinetic' duties.

At its most basic, area denial aims to deter insurgent/terrorist attack in localities where government presence is too weak to provide security to the civilian population. It involves selectively arming members of the population, and having them serve as a tripwire that would warn security forces of insurgent movement within a given area. Reaction forces can then be mobilised to deal with the insurgents, even as they are being engaged in combat by pro-government militiamen.

The historical record on area denial is mixed: Such missions are relatively cheap and can seriously hinder insurgent/terrorist activity, but they are not in themselves transformative. Imperfections in their implementation pose the biggest obstacle to achieving strategic results. Lack of dedicated and talented trainers is one such imperfection, poor vetting of local volunteers is another, while distrust and lukewarm support from both political and senior military leaders is a third. From two examples, French civic action teams in Algeria and American Special Forces in Vietnam, it appears that the most important lesson of area denial missions is: Have moderate expectations.[6]

FRENCH CIVIC ACTION IN ALGERIA

Until 1962, Algeria was part of France, but was governed harshly, as though it were a colony. In 1954, a rebellion broke out, aiming to achieve Algerian independence. By 1955 the French Army had begun to raise civic action teams with the objective of winning over the rural population, who were highly susceptible to separatist subversion. Called Special Administrative Sections (SAS—not to be confused with the British Special Air Service), these teams consisted of a captain or lieutenant, his deputy, a secretary and an interpreter. They were protected by a force of between 30-50 locally-recruited bodyguards, called *harkis*. The SAS teams lived and worked full-time in remote villages, providing education, medical assistance and agronomical advice and mediating disputes.[7]

Given the level of expertise demanded by their profession, French SAS officers in Algeria needed to be both highly educated and highly committed to their job, despite harsh living conditions. This is a rare combination to find in any era, and the 1950s were not much different. Many SAS officers inspired loyalty among their *harki* bodyguards, and motivated them to chase out separatist guerrillas from their localities. Others were hampered by a lack of personal charisma, inadequate language skills and the poor opinion which French military commanders had of the whole SAS concept. They were, thus, not able to obtain food or accommodation allowances

---

[6] Clegg, Will, 'Irregular Forces in Counterinsurgency Warfare', *Security Challenges*, Vol. 5, No. 3 (2009), p. 8.

[7] Sutton, Keith, 'Army Administration Tensions over Algeria's Centres de Regroupment, 1954-1962', *British Journal of Middle Eastern Studies*, Vol. 26, No. 2 (1999), p. 261.

for their men, or raise their wages. As a result, some *harkis* were not interested in fighting for the French government.

By 1962, the French Army had implanted 400 SAS teams across Algeria. Each was responsible for administering three to four large villages, with a total population of between 10,000 and 15,000. Their work threatened the political legitimacy of the separatist movement, because it showed that Frenchmen and Algerians could live alongside each other and work for a common future. Separatist assassination squads, therefore, singled out SAS personnel for elimination: Over six years, 73 officers, 33 sub-officers and 600 *harkis* were killed by the rebels, making the SAS the most threatened component of the French Army. Their access to the local population had made SAS officers excellent collectors of grassroots intelligence. However, even in the localities where they were highly successful in winning over the local population, they were only able to forge bonds based on interpersonal loyalty. Their *harki* bodyguards were not interested in fighting to ensure that Algeria remained a part of France, but merely to protect the officers from separatist assassins. Eventually, the SAS experiment proved to be a valiant effort at winning hearts and minds, but it did not affect the course of the war.[8]

US SPECIAL FORCES IN VIETNAM

During the Vietnam War, the US Army Special Forces (commonly known as 'Green Berets') raised anti-guerrilla units among Montagnard tribesmen. The Montagnards were an ethnic minority in South Vietnam, who, although hostile to the insurgents, had also been badly treated by the government. Consequently, their loyalty to the South Vietnamese state was non-existent, and they fought entirely for the wages or rations that they received. The tribesmen

[8]Orwin, Etan M, 'Squad leaders today, village leaders tomorrow: Muslim auxiliaries and tactical politics in Algeria, 1956-1962', *Small Wars & Insurgencies*, Vol. 23, No. 2 (2012), pp. 336-337.

were organised into two types of unit: Hamlet militias and strike forces. The former were part-time units that received no pay and only minimal arms training. Their job was to keep their own village clean of insurgents and if any approached, to call for help from the nearest strike force. The strike force members were full-time soldiers. Their job was to fortify villages against insurgent attack and to provide a quick response whenever an attack occurred. US Special Forces soldiers were responsible for training both types of unit, and for liaison with the regular US and South Vietnamese military forces.

The Montagnard recruitment effort, known as the Civilian Irregular Defense Group (CIDG) programme, was hamstrung by lack of political support and poor vetting procedures. South Vietnamese leaders were unwilling to provide weapons to the minority community, fearful that it would rise up against the state. Owing to the long history of official discrimination against the Montagnards, communist insurgents were easily able to subvert members of the hamlet militias and strike forces. In one attack after another on US Special Forces camps (which also served as training centres for CIDG units), the insurgents used saboteurs and spies to carry out highly effective precision strikes.

Even so, attacks on Special Forces/CIDG camps were costly for the insurgents, as those Montagnard tribesmen who had not been subverted, tended to fight well in fixed defensive combat. In cross-country operations, their effectiveness was crucially dependent on the level of support they received from regular US and South Vietnamese forces. Their main contribution to the counter-insurgency effort was in providing terrain knowledge that helped regular forces chase the insurgents across rough ground. This is an important point—it corroborates the larger argument of this chapter, that the activities of special operations forces need to be linked to broader offensive measures, spearheaded by conventional units. If they operate in isolation, SOF have limited value.

The CIDG example proves the point: When it came to their own combat performance under heavy fire, the Montagnards were not very successful. Communist insurgents proved more than a match for them, being battle-hardened fighters accustomed to going up against the Americans in close quarter battle. Yet, despite its limited military utility, the US found the Montagnard programme to be reasonably worthwhile. Each Montagnard soldier cost an average of 71 per cent of the amount that was needed to train an American soldier, but his maintenance costs in the field were much lower at 16 per cent.[9] This meant that in the long run, using Montagnards to keep the insurgents bleeding was an attractive option. As public support for the Vietnam War dwindled in the US, the CIDG programme came to be viewed as a way of waging counter-insurgency on the cheap. Instead of large numbers of US infantrymen and Marines, all that was required was a few Green Beret trainers and a moderate amount of logistical support.

Area Dominance Missions

Area dominance missions are based on recognition of a single principle: Popular support in an insurgency goes to the side that is locally powerful, not the side that people actually sympathise with. When security forces physically control a locality that is sympathetic towards insurgents/terrorists, they can still gather intelligence, despite public hostility, by regulating the pace of daily life. For example, an informer grid can be created by imposing movement controls on the general population and selectively easing these for people who agree to provide information.[10]

---

[9] Stanton, *Green Berets at War*, p. 183.

[10] An example being Israel's continuing recruitment of Palestinian informers in the West Bank and Gaza. See Cohen, Hillel and Dudai, Ron, 'Human Rights Dilemmas in Using Informers to Combat Terrorism: The Israeli-Palestinian Case', *Terrorism and Political Violence*, Vol. 17, Nos.1-2 (2005), pp. 232-233.

Special operations that aim to wrest control of a locality from insurgents must necessarily be coordinated with larger ground offensives. By themselves, these operations can be death traps for the personnel involved but when employed in tandem with conventional forces, they can dramatically alter the military balance of power. An example of this transformative effect can be found in the Yemeni campaign waged by the Egyptian Army in the 1960s. The Egyptians wanted to defeat an Islamist insurgency in North Yemen that was being covertly supported by Saudi Arabia. Since the terrain was extremely rugged and mostly un-motorable, Egyptian troops faced serious difficulty merely while entering rebel-held territory, never mind occupying it. For four months, they tried vainly to dominate the rebel strongholds, but were denied space and time to make a breakthrough. Then, in early 1963 they carried out a coordinated three-week offensive that smashed the rebel defences.

The key to this sudden reversal of fortunes was a paratrooper battalion that was airdropped deep inside the rebel heartland and tasked to prepare a dirt runway. Using this runway as an airhead, the Egyptians then landed a full airborne brigade behind enemy lines. Even as the rebels struggled to understand what was going on to their rear, armoured columns supported by air cover launched a frontal attack on rebel-held roads. Clearing the blockades that had been meant to stop them, the columns raced into Islamist areas faster than the insurgents could organise a defence. By the end of the offensive, the insurgents' combat capability had been severely degraded, partly because they had made the mistake of trying to fight positional battles against a fast-moving enemy.

Through using armour and paratroopers in an unorthodox manner as self-sufficient forces, the Egyptians had adopted a special operations approach in the counter-insurgency campaign.[11]

[11] Witty, David M, 'A Regular Army in Counterinsurgency Operations: Egypt in North Yemen, 1962-1967', *Journal of Military History*, Vol. 65, No. 2 (2001), p. 414.

The rapidity of their movement cancelled the traditional advantage that guerrillas enjoy when confronted by superior forces—the ability to disperse across time and space. Instead, a certain symmetry appeared in the two sides' tactics—each was fighting in the form of small, autonomous units scattered across the country. The result was a foregone conclusion: As the Harvard scholar Ivan Arreguin-Toft has predicted, whenever government forces adopt the insurgents' own tactics, the government will gain a crucial advantage.[12] Thus, were the Egyptians able to dominate rebel strongholds in North Yemen.

For its part, the US learned the value of emulating an irregular adversary's own tactics in Latin America, where it sponsored death squads to eliminate leftist rebels and drug lords. During the 1980s and '90s, US Special Forces trained Colombian militiamen to hunt down narco-traffickers, using a variety of means that American law enforcement personnel were officially forbidden to use. The 'Salvador Model', as the use of death squads has come to be known, was also applied in post-Saddam Iraq during the early phases of the insurgency, in order to wipe out the rebel leadership. It is another matter that the Americans targeted the wrong people: Instead of foreign *jihadists*, the death squads set up by the US military killed mid-ranking Ba'ath Party officials who had little to do with the insurgency.[13] Yet, as the French Army previously showed in Algeria, area dominance is potentially the most powerful instrument that any government has, in its efforts to defeat an insurgency.

THE CHALLE PLAN: SPECIAL OPERATIONS IN ALGERIA

In Algeria, the French experimented with highly innovative tactics that delivered good battlefield results. Unfortunately, these came too late to affect the political mood of defeatism that was prevailing in

---

[12] Arreguin-Toft, Ivan, 'How the Weak Win Wars: A Theory of Asymmetric Conflict', *International Security*, Vol. 26, No. 1 (2001), p. 110.

[13] Davies, Nicolas JS, 'Evidence of an American Dirty War in Iraq', *Peace Review: A Journal of Social Justice*, Vol. 19, No. 3 (2007), pp. 435-437.

France—a mood which allowed Algerian separatists to win the war for independence despite having been militarily thrashed. Perhaps the most important of these innovations was the use of special operations units to track and eliminate insurgent bands in the Algerian countryside.

In February 1959, the French Army launched a series of offensives that became known as the Challe Plan after its author, General Maurice Challe, one of the key military leaders during the Algerian War. The offensives lasted until April 1960, by which time they had cumulatively killed an estimated 45,800 insurgents, led to the capture of 13,500 more and the seizure of 26,150 weapons. Over slightly more than a year, the French degraded the total combat power of the separatists (who enjoyed sanctuary in neighbouring Morocco and Tunisia) by 50 per cent.[14]

The Challe Plan was highly effective because by 1959, the separatists had been driven out of Algerian cities. Determined police action against bomb-making networks had brought down the threat of urban terrorism to negligible levels, while intensified border surveillance made surreptitious crossings from Tunisia and Morocco highly dangerous. The counter-insurgency campaign became a numbers game, with the French Army only having to kill or capture insurgents at a faster rate than at which they could enter the country from outside. This was where the Challe Plan came in.

First, the plan divided Algeria into a counter-insurgency grid, composed of zones that were further sub-divided into sectors. Each sector was controlled by a battalion that provided security to the local population and gathered intelligence. These stationary troops were known as 'Quadrillage' troops. Their deployment was especially dense in areas that were known hotbeds of insurgent activity. Then there were the mobile reserve troops, known as Reserves

[14] Griffin, Christopher, 'Major Combat Operations and Counterinsurgency Warfare: Plan Challe in Algeria, 1959-1960', *Security Studies*, Vol. 19, No. 3 (2010), pp. 571-572.

Generales (RG). These were either paratroopers or Foreign Legion units that had been provided a tactical airlift capability. Their job was to respond swiftly to sightings of large insurgent bands and eliminate them. Finally, the most important component of the Challe Plan was the special operations element, called Commandos de Chasse. The Commandos were elite troops organised into company-sized units, who were tasked with 'tagging' a comparably sized insurgent band (known as a *katiba*) and pursuing it relentlessly across the country. The Commandos de Chasse were not subject to any jurisdictional boundaries; they could traverse the length and breadth of Algeria for as long as was needed to destroy a particular *katiba.*

Key to the Commandos' effectiveness was the intelligence that they received from SAS officers in remote areas, and from Quadrillage troops. When pursing an insurgent *katiba*, the Commandos were fed a constant stream of information relevant to their tactical difficulties. For this purpose, French Army used aircraft to conduct overhead surveillance of operational areas and track the movement of fleeing *katibas*. Sometimes, these *katibas* were manouevered into free-fire zones away from human habitation, where they could be neutralised by aerial bombardment. At other times, they were chased to waiting RG troops who had been helicoptered into an ambush point further along their escape route.

Special operations carried out by the Commandos de Chasse were successful because they were conducted within territory that was still part of France i.e., not across an international border. The Commandos could, thus, call upon airmobile reinforcements whenever they were faced with superior enemy numbers. Furthermore, they had the benefit of at least some local support, even in areas where the insurgents enjoyed widespread sympathy. Some of the Commandos were themselves Algerian Muslim soldiers. Of the seventy-four Commando de Chasse units that were created by early 1959, most had some Muslim soldiers, who were more adept at gathering intelligence from the local population. Six units were entirely composed of Muslims, with the exception of officers and warrant officers, who were French. These six units between them killed 621 insurgents for a loss of thirty-seven of their own men.[15]

As the Challe Plan showed, the use of special operations within friendly or semi-controlled environments allows them to deliver good results. Unlike area denial missions, area dominance missions involve time-bound pursuit of guerrillas across rough terrain, in coordination with regular forces, with air power providing much-needed fire and logistics support. In addition, new troops can also be inducted into an operational zone to hit the enemy while he is still off-balance and struggling to recover from initial attacks. Thus, area dominance missions can be far more effective than area denial at degrading an irregular adversary's combat potential.

Area Penetration Missions

Even area dominance missions, however, cannot completely defeat an insurgency. They can degrade the insurgents' attack capability within a country, but if the insurgency has a cross-border sanctuary, it will survive this. The intention to carry out attacks on the victim state will remain, even if the capability is weakened. To comprehensively defeat an insurgency, therefore, it is necessary to disabuse the insurgents of any idea that they will gain an advantage by continuing hostilities. Unfortunately, the history of special operations in irregular warfare suggests that changing an adversary's intention is next to impossible. All that can be done (perhaps) is to redirect the adversary's anger onto a more proximate target, even if that target is a patron state that supports the insurgency.

When it comes to direct action, cross-border special operations are high-risk, low-return ventures. The only exception to this

[15] Gortzak, Yoav, 'Using Indigenous Forces in Counterinsurgency Operations: The French in Algeria, 1954-1962', *Journal of Strategic Studies*, Vol. 32, No. 2 (2009), pp. 326-328.

general principle is when the special operations are carried out very selectively, against protected targets located deep within enemy territory. In such scenarios, area penetration special operations retain their high-risk character, but they also offer much higher returns that justify the risk. By exposing the vulnerability of a target state to an extent that was previously unimaginable, they can influence its behaviour positively. A state sponsor of terrorism, for instance, might be prepared for attacks on its outlying border posts, but not its army or intelligence headquarters.

LESSONS FROM THE CASE OF SOG-MACV

During the Vietnam War, the US initiated a fascinating experiment in covert operations, one which is directly relevant to India's current problem with cross-border terrorism. The communist state of North Vietnam was supporting insurgents in South Vietnam, a US ally. North Vietnam did this while consistently disclaiming that it was instigating violence. Realising that its nuclear arsenal was no deterrent to proxy warfare, the US decided to retaliate with a proxy war of its own. There was only one problem, though—it had no reliable proxies to use against the North.

Operatives of the Central Intelligence Agency (CIA) tried in vain to raise agents for covert action, but practical difficulties caused by the harsh counterintelligence system in North Vietnam made this a non-starter. In frustration, the US defence secretary authorised the Pentagon to create a special unit for covert operations in the North, using military personnel and resources. On 24 January 1964, the Studies and Observations Group—Military Assistance Command Vietnam (SOG-MACV) was created. It consisted of personnel from the US Army, Navy, Air Force and Marines, but was financed by the Navy. SOG-MACV had four tasks:

- To insert agents into North Vietnam for a campaign of sabotage and subversion,

- To conduct 'black' propaganda that would feed the North Vietnamese government's paranoia about internal threats,
- To conduct maritime raids on coastal targets in the North and
- To gather intelligence on North Vietnamese activities along the Ho Chi Minh Trail, with a view towards disrupting them.

The Ho Chi Minh Trail was the main logistical lifeline between North Vietnam and the communist insurgency in the South. It bypassed the heavily guarded border between the two states, and instead snaked southwards through the neutral territory of neighbouring Laos and Cambodia. The governments of these countries turned a blind eye to North Vietnamese activities on their soil. SOG-MACV was, therefore, tasked to conduct deniable operations in Laos and Cambodia to monitor the Trail and attack supply convoys that travelled down it, while simultaneously disrupting rear area security in North Vietnam itself. The unit's supporters in the Pentagon hierarchy assumed that if confronted with the threat of a US-supported insurgency on its own territory, North Vietnam would back off from supporting the insurgency in South Vietnam.

SOG-MACV's formal mandate was defined in a document called Operational Plan (OPLAN) 34A. It proposed seventy-two different types of action that could be taken to destabilise North Vietnam. Of these, thirty-three were eventually implemented. The remainder were held in abeyance because the US officials were concerned that if covert operations were too successful, they might provoke North Vietnam into escalating its support to the South Vietnamese communists. Even worse was the possibility that China might intervene on the side of North Vietnam. In worrying about these scenarios, US planners got ahead of themselves: They forgot that the CIA's experience in covert operations had shown that success in denied territories is hard to come by. Rather than pre-emptively restraining itself, the US should first have gone all-out to destabilise North Vietnam, if it wanted to scare the regime.[16] As things turned out, the North Vietnamese government was hardly impressed by vague American threats of supporting a counter-proxy war, and easily discredited these.

Almost all of the roughly 500 agents who were sent into North Vietnam by SOG-MACV between 1964 and 1968 were arrested and 'turned' in captivity[17]. The destabilisation programme failed because of poor operational security: Agents selected to work in North Vietnam were recruited from diaspora communities in Laos and Cambodia, where North Vietnamese intelligence already had a strong presence. It is, thus, likely that several were double agents from the very beginning. Even those who were genuinely committed to carrying out sabotage in the North, were probably betrayed by communist spies in the SOG-MACV linguistics cell, which was manned by South Vietnamese soldiers. In hindsight, bureaucratic pressure to generate quick results damaged the programme beyond recovery. Upon realising that all SOG-MACV agents in North Vietnam were actually working on the enemy's behalf, a CIA officer told the unit chief: 'What do you want to tell Ho Chi Minh [North Vietnam's leader]? Your teams are double agents and I can send Ho the message through them'.[18]

If SOG operations in North Vietnam were a disaster, the unit still enjoyed tremendous success in its operations against the Ho Chi Minh Trail. Throughout the war, roughly fifty SOG men operated

---

[16]Schultz, Richard H, Jr, *The Secret War Against Hanoi: Kennedy's and Johnson's Use of Spies, Saboteurs, and Covert Warriors in North Vietnam* (New York: HarperCollins, 1999), pp. 37-40.

[17]Andrade, Dale, 'The Vietnam War's Secret Side', *International Journal of Intelligence and CounterIntelligence*, Vol. 14, No. 4 (2001), pp. 582-585.

[18]Crist, David, *The Twilight War: The Secret History of America's Thirty-Year Conflict with Iran* (New York: Penguin, 2012), p. 58.

in Laos and Cambodia at any one time. In comparison, the North Vietnamese Army assigned 30,000 men to guard the Trail against attack and hunt down SOG teams. This means that for every SOG man sent into the field, a battalion-sized force (600 enemy soldiers) was tied down in defensive duties. Equally impressive were SOG direct actions along the Trail. The unit achieved a kill ratio of 1:100, as compared to the average ratio of 1:15 for the rest of the US military forces deployed in Vietnam.[19] The much greater lethality of its operations was derived from two factors: Liberal use of air power to strafe and bomb enemy troop concentrations in Laos and Cambodia, and the general sense of complacency among enemy troops, the further they were from a combat zone.

The case of SOG-MACV provides two lessons regarding the use of area penetration missions. First, weak operational security when compounded with political pressure to generate quick results, can doom missions to failure. Second, area penetration missions can be devastating to both enemy morale and infrastructure, provided they occur deep inside enemy territory. Initially, SOG spent a lot of time carrying out shallow-penetration ambushes and raids on parts of the Ho Chi Minh Trail that lay close to the border. The North Vietnamese responded to these by setting up dedicated combat tracker teams, and establishing a network of 'Landing Zone Watchers' and 'Trail Watchers'. The Watchers would observe SOG teams trying to enter Laos and Cambodia and alert the tracker teams, who would then use their superior knowledge of local terrain to set up ambushes. SOG had no reply to these countermeasures, except to carry out attacks further inside Laotian and Cambodian territory, where vigilance levels were lower. These attacks in turn, were highly successful because they were directed against troops who were

[19] Plaster, John L, *SOG: The Secret Wars of America's Commandos in Vietnam* (London: Onyx, 1998), p. 267.

unprepared to undertake defensive action. India should take note of this: If Indian Special Forces are ever deployed against targets in Pakistan, they should hit sites that are far from the border, because such attacks will be relatively unopposed once launched. The challenge will be in arranging logistics to secretly transport a raiding force deep inside enemy territory.

Layering of Special Operations Forces

The US military has perhaps the most elaborate system of planning and supporting special operations anywhere in the world. To ensure bureaucratic and financial autonomy as well as a dedicated career stream for special operations personnel, in 1987 it established Special Operations Command (SOCOM). With a strength of 46,000 all ranks in the 1990s, SOCOM was a supporting command that was tasked to provide assistance to the various regional commands. The bulk of its units were 'white' or 'vanilla' SOF, such as the US Army Special Forces Groups. However, a 1300-strong sub-command, called the Joint Special Operations Command, consisted of 'black' units that were meant to carry out highly sensitive missions, mostly relating to counterterrorism.[20]

The 'white' versus 'black' distinction in the US special operations community is interesting, because it reflects a degree of expertise that India would do well to adopt. The 'white' units are ideal for area denial missions, while the 'black' units are suited for area penetration missions. US Green Berets for instance, are trained to build a rapport with tribal chieftains and guerrilla leaders in foreign countries, and also with military leaders, for the purpose of pushing forward American strategic interests. They are 'armed anthropologists' who have to be culturally sensitive and extremely patient while operating in unfamiliar environments. 'Black' units on the other hand, such as

[20]Waller, Douglas C, *The Commandos: The Inside Story of America's Secret Soldiers* (New York: Dell, 1995), p. 33.

Delta Force, SEAL Team 6 and 75th Ranger Regiment, have neither time nor respect for native customs. They are hard-hitting outfits meant for kinetic action—get in there, kill the bad guys and anyone protecting them, and get out.

For much of the 1990s, JSOC had to invent work for itself. Counterterrorism was not a policy priority as far as the top-most American decision-makers were concerned, so US special operations forces were deployed on training missions and humanitarian operations. It was only after 9/11 that the 'black' units really received political backing to do the job they had been established for. Soon after the attacks on the World Trade Center and the Pentagon, US defence secretary Donald Rumsfeld became the chief patron of JSOC, upgrading its status from a supporting command to a supported command. Now, the theatre commands needed to defer to JSOC requests for assistance, whenever these were made. A parallel army was created outside the established military hierarchy, in which private security companies like Blackwater played a crucial role in gathering intelligence and training Afghan forces for cross-border raids inside Pakistan. The US exploited Pakistani vulnerability after the 2005 earthquake to flood the country with intelligence operatives, ostensibly working as aid activists.

India does not have resources that are anywhere near comparable to those of the US. However, it can recognise that special operations require a dedicated funding stream and career track, and create a tri-services command exclusively for such operations. The US was able to kill Osama bin Laden because it had spent over twenty years investing in the tactics and technologies needed to carry out long-range area penetration missions. Equally important, JSOC was able to navigate the corridors of power in Washington and gain access to top decision-makers.

After Barack Obama was elected president, JSOC impressed him by its sharpshooting skills—in a brilliantly coordinated hit, three snipers from SEAL Team 6 killed three Somali pirates who were holding an American hostage. The pirates and hostage had been aboard a small dingy in choppy seas at the time, and the SEALs' accuracy was indeed spectacular.[21] The accolades that Obama received as a result of this rescue convinced him that the political payoff from a successful special operation could be worth the risks that might ensue from its failure. Indian politicians would do well to draw the same lesson, given that terrorism remains a far more serious threat to India than to the US.

[21] Scahill, Jeremy, *Dirty Wars: The World is a Battlefield* (London: Serpent's Tail, 2013), pp. 275-276.

2

Intelligence

While studying the use of special operations against irregular adversaries, it is necessary to consider the role played by intelligence. This chapter provides examples that illustrate the centrality of intelligence to special operations, and explains the difficulties associated with gathering such intelligence. Drawing from the American and British experience of counterterrorism, it will conclude by providing a tentative suggestion of how the difficulties can be surmounted.

There are three factors that are needed to carry out special operations against terrorist groups: Intelligence, intelligence and intelligence. This is not a joke. Intelligence has a triple significance in determining the outcome of any cross-border special operation. First, it is vital in identifying and locating targets. Second, it helps maximise the possibility of success when a target is actually engaged, by enabling surgical neutralisation of defensive positions. Finally, it minimises the risk of failure once the enemy has recovered from his initial paralysis and begun to organise a response, since it permits rapid exfiltration of the attacking force from the engagement area.

Without specific and timely intelligence, it is impossible for governments to retaliate to terrorist outrages. Unfortunately, very rarely is intelligence both specific and timely. In the immediate aftermath of a major attack, the first question that puzzles intelligence agencies is, who did this?[1] Various theories abound, but in the absence of strong leads (which typically do not become available until much later) these theories are basically guesswork. Even once a group has been identified as the likely perpetrator, the next logical question becomes, why would this group carry out such a provocative action? What did it hope to achieve? As precious days are wasted, the political window for taking punitive measures becomes smaller, as international sympathy begins to wane for the victim state. Even more insidious than the loss of international support is the weakening of political will domestically for exacting vengeance upon the perpetrators.

Prevention Through Retaliation

One disgraceful example is the American failure to respond to a 1983 bombing in Lebanon. The bombing killed 141 US Marines and provoked a surge of public and political outrage within the United States. Yet, even as President Ronald Reagan weighed up the pros and cons of retaliation, his defence secretary, Caspar Weinberger, sabotaged any move towards hitting back. Weinberger was fearful that responding to the attack would draw the US into the Lebanese Civil War, and lead to another Vietnam-style embroilment. He, thus, misused his authority as head of the US defence establishment to countermand verbal orders that the president issued for carrying out

[1] Cieave, Michelle Van, 'Strategic Counterintelligence: What Is It and What Should We Do About It?', *Studies in Intelligence*, Vol. 51, No. 2 (2006), https://www.cia.gov/library/center-for-the-study-of-intelligence/csi-publications/csi-studies/studies/vol51no2/strategic-counterintelligence.html, 8 No· mber 2013.

punitive air strikes on terrorist camps in Lebanon. When confronted with the defence secretary's insubordination, Reagan demurred from overruling him, and instead persuaded himself that his initial orders for reprisals had been too hasty.[2]

The matter did not end there; when yet another bombing hit the US embassy in Beirut the following year, Reagan was much less inclined to authorise a hard response. Having procrastinated after the attack on the Marines, he found it politically easy to justify inaction in this case as well. What he did not realise was that by not retaliating to the deaths of US servicemen and diplomats, he was inadvertently giving credibility to a narrative among international terrorists, that the US was a soft state. Fearful of losing yet more lives in unnecessary wars overseas, it was ready to turn the other cheek whenever its citizens were attacked. This narrative was bought into by Osama bin Laden, who was inspired by the 1983 and 1984 bombings and used them as a model to plan even more destructive attacks against the US, in order to make a name for himself in the international terrorism fraternity.

All the subsequent bloodshed could have been avoided if the US intelligence community had been able to capitalise on the initial resolve shown by the political establishment, and identified targets for immediate retaliation. Weinberger's damage was done in the time lag between first news of the Marine bombing arriving and the acquisition of hard evidence regarding the perpetrators' identities and locations. During this period, concerns over collateral damage cropped up, US strategic interests in the Middle East were cited to justify doing nothing, and Reagan began backing away from his initial hard line reaction, not wanting to lose the support of doves in his cabinet. From this example, Indian intelligence agencies should note that failure to swiftly affix culpability for a major terrorist attack

[2] Crist, *The Twilight War*, pp. 146-147.

will play into the hands of the perpetrators, as well as those within India and abroad who advocate passiveness.

Intelligence provides targets for retaliation and in rare cases, for pre-emption. During the 1970 Dawson's Field crisis, Palestinian hijackers threatened to specifically harm Israeli hostages if their demands were not met. Intelligence operatives then traced the relatives of senior Palestinian leaders who were in contact with the hijackers, and arrested dozens of them. Israel sent a message through Arab intermediaries, warning that treatment of the relatives would be reciprocal in case any harm came to the Israeli passengers.[3] The tactic worked because the terrorist leadership was closely connected to the on-ground cadres who were actually involved in the hijacking, and was able to restrain them. Without intelligence on the perpetrators' identity and their exploitable weaknesses, it would not have been possible for Israel to deter the hijackers from killing its citizens.

An even more successful example was Soviet use of intelligence to resolve a hostage crisis in Lebanon. On 30 September 1985, four Soviet diplomats were seized by a faction of Hizballah. One of the diplomats was executed shortly thereafter. Soviet operatives, probably from the Alpha counterterrorist unit of the Spetsnaz, swiftly retaliated by snatching three male relatives of a Hizballah leader. They cut off the testicles of one, stuffed them into his mouth and sawed off his head. The head was returned to the family home with a note, stating that if the remaining Soviet hostages were not released, the testicles of the other two relatives would follow, inside the heads of their wives. The diplomats were freed and although dozens of Western nationals continued to be taken hostage by Hizballah over the following years, the group never dared target the Soviets again.

Unfortunately, it is very rare that intelligence can be so precise as

---

[3] [Anonymous] 'After Munich: Israel Debates the Response', *Journal of Palestine Studies*, Vol. 2, No. 2 (1973), p. 146.

to be useful during a crisis. Oftentimes, the most important details regarding perpetrator identity and vulnerability only become available well after a crisis has passed and done whatever damage it could. It is a delicate task to optimise intelligence for use in special operations, made all the more complicated by the fundamental difficulties of collecting information in hostile environments. Normally, this job could be left to technical collection systems, but with special operations, human agents come into play.

A Resource-intensive Activity with Low Success Rates

The first point which needs to be understood about intelligence collection from human sources is that it is time-consuming. Even domestic intelligence, which is much easier to collect than foreign intelligence, has a high wastage rate in terms of man-hours spent cultivating potential sources. The fact is that most intelligence handlers have no special gift for recruiting informers—they need to wait for an opportunity to 'pitch' the subject when s/he is emotionally vulnerable. However, pressure to deliver fast results can lead to shortcuts being taken; instead of manipulating a potential source into agreeing to work as a spy, they coerce or blackmail him/her. This tactic is especially popular with police and counterintelligence services, but can backfire badly if tried in a foreign setting.

A source who has been blackmailed into spying could try to regain control of the situation by 'doubling' against his/her handlers. The ever-present risk of betrayal means that information from a single source, no matter how sensational or crucial, usually needs to be corroborated by at least two other sources before it is treated as authentic. This slows the analytical process, which is not a problem during peacetime, but can be fatal in wartime or in the unique context of counterterrorist operations.

In Russia, the Federal Security Service (FSB) faces various

difficulties in generating actionable intelligence, due to its own in-house restrictions. Its handling officers tend to be very cautious about information received from Chechen sources, for fear that these might have 'doubled' on them. Although such caution is justifiable, it requires that Russian counterterrorist operations depend on electronic intercepts, which pose another kind of problem: Lack of sufficiently qualified translators. Most intercepts cannot be transcribed because the FSB has very few Chechen-speaking officers.[4]

Special operations require a peculiar kind of intelligence: One that is acquired in hostile territory but can also be constantly updated upon request. It is, thus, a mix of data obtained through reconnaissance and through surveillance. Yet, as one veteran American official has noted, the two types of intelligence system, reconnaissance and surveillance, rarely work alongside each other. One is used for supporting wartime military operations while the other is used for keeping the peace. When counterterrorist intelligence is collected through a reconnaissance-oriented system it becomes outdated very fast as events on the ground overtake it. To be truly effective, such intelligence must be collected through a surveillance-based system that can 'tag' a specific target and keep it under constant observation despite all attempts at evasion and concealment.[5] Achieving such a degree of granularity in intelligence coverage is extremely costly. Even in domestic surveillance of terrorist suspects, a 24-hour watch pattern can require as many as sixty operatives to be assigned to a single target.[6]

---

[4] Kramer, Mark, 'Guerrilla Warfare, Counterinsurgency and Terrorism in the North Caucasus: The Military Dimension of the Russo-Chechen Conflict', *Europe-Asia Studies*, Vol. 57, No. 2 (2005), pp. 249-250.

[5] Cline, Lawrence S, 'Special Operations and the Intelligence System', *International Journal of Intelligence and CounterIntelligence*, Vol. 18, No. 4 (2005), p. 581.

[6] Hewitt, Steve, *Snitch: A History of the Modern Intelligence Informer* (New York: Continuum, 2010), p. 123.

Terrorist Counterintelligence

Another serious problem is that terrorists take every precaution against intelligence penetration. Not only do these precautions include routine 'passive' measures, such as communications security and stringent personnel vetting, but they also include opportunistic 'active' measures, such as doubling police informers and using them to pass on disinformation. Together, passive and active counterintelligence make penetration of terrorist groups extremely difficult, and even in those cases when it is possible, extremely risky. Handlers can never be sure when their prized agent might be turned against them by the terrorist organisation's mole-hunting unit, and used to lure them into a trap. This is what happened to a Europe-based Mossad officer in 2000 when he attempted to conduct a false-flag recruitment by posing as a handler from a third country's intelligence agency. Hizballah, the group that was being targeted for penetration, swiftly realised what he was up to and lured him to Beirut, where its operatives grabbed him.[7]

All terrorist groups practice some kind of counterintelligence. Al Qaeda preferred to rely on passive measures, sensitising its operatives about US signals intelligence capabilities. Its operational doctrine urged that each draft plan for carrying out a terrorist attack should include a built-in security component, before being submitted for internal review.[8] However, the group also tried to penetrate the US intelligence community by sending sleeper agents to volunteer for translation work. In 2009, it scored an impressive success when one of its double agents killed several CIA officials in a

---

[7] Wege, Carl Anthony, 'Hizballah's Counterintelligence Apparatus', *International Journal of Intelligence and CounterIntelligence*, Vol. 25, No. 4 (2012), pp. 773-778.

[8] Ilardi, Gaetano Joe, 'Al Qaeda's Counterintelligence Doctrine: The Pursuit of Operational Certainty and Control', *International Journal of Intelligence and CounterIntelligence*, Vol. 22, No. 2 (2009), p. 249.

suicide bombing in Afghanistan. The agent had so completely won the trust of his handlers that he was allowed to meet them without undergoing security screening—a fatal mistake.[9]

Like Al Qaeda, the Irish Republic Army (IRA) used a combination of passive and active measures to protect its secrets. It organised its attack cells into four to five men active service units (ASUs) that would plan operations on a semi-autonomous basis. Yet, unlike Al Qaeda, it was unsuccessful at preventing intelligence penetration of its ranks, because it operated in territory controlled by its main opponent—the British government. Up to 80 per cent of its operations were disrupted on the basis of intelligence collected through both human and technical means. However, when the IRA switched from passive to active counterintelligence measures, it was able to spring a few nasty surprises because the British were blissfully unaware of their own security weaknesses. The most spectacular of these occurred in 2002, when IRA operatives penetrated a police Special Branch headquarters in Northern Ireland and stole compact discs containing the names of over 250 police informers and their handlers.[10] Never before had the British security establishment suffered such a grave loss of sensitive data.

Both Al Qaeda and the IRA pale in comparison to Hizballah, a Lebanon-based militant group which is considered by some scholars to be the most sophisticated terrorist group in the world. Hizballah has a transcontinental network of sleeper agents, mostly recruited through family and clan connections in the Lebanese diaspora. Western efforts to get African and Latin American security services,

---

[9] Oppel, Richard A, Jr, Mark Mazzetti and Souad Mekhennet, 'Attacker in Afghanistan Was a Double Agent', *New York Times*, 4 January 2010, http://www.nytimes.com/2010/01/05/world/asia/05cia.html?_r=0, 8 November 2013.

[10] Ilardi, Gaetano Joe, 'Irish Republican Army Counterintelligence', *International Journal of Intelligence and CounterIntelligence*, Vol. 23, No. 1 (2010), p. 20.

in whose jurisdiction the group has a strong presence, to cooperate in counterterrorism investigations have yielded poor results. So powerful is Hizballah's active counterintelligence wing that it has consistently been able to frustrate inter-governmental efforts to monitor the group. One chilling example comes from a South American country, where US Federal Bureau of Investigation (FBI) officials had been planning to set up a surveillance mission. Their arrival in the country was supposed to be top-secret, with foreknowledge being confined only to the senior-most officials of the local partnering service. Yet, even before the FBI surveillance team had entered the airport building after deplaning, their pagers went off. They immediately telephoned their New York office, only to be told that a photograph of them getting off the plane had arrived by fax, while they were still walking across the runway.[11]

Even in passive measures, Hizballah has been able to compartmentalise its operations so well that it is very difficult for Western or Israeli intelligence agencies to gain advance warning of an attack. Typically, most of its operatives have a wide circle of 'social contacts', i.e., friends, relatives and neighbours who could be tapped for information. However, the actual planning of an attack is limited to a much smaller circle of 'operational contacts'—people who do not socialise with one another, but are drawn from different branches of the organisation and assembled together to achieve a one-off objective. Once their task has been completed, these individuals disperse back to their parent branch and do not have much contact unless they are assembled again for another operation. This means that a single informer cannot provide a continuous stream of reports on a single Hizballah network, as the component parts of that network are constantly switched around, re-tasked

[11] Constanza, William, 'Hizballah and Its Mission in Latin America', *Studies in Conflict & Terrorism*, Vol. 35, No. 3 (2012), pp. 197-198.

and interchanged. Instead of a few high-level sources, intelligence agencies need to use many low-level sources just to keep track of Hizballah activity.[12]

Need To Combine Intelligence With Operations

Given how effective terrorist groups can be at keeping secrets, intelligence agencies rarely penetrate them to the extent required to support special operations. Instead, such operations must support themselves, by creating a loop between direct action missions and the information gathered thereby, thus allowing even more missions to be carried out.[13] Ideally, special operations would be initiated according to an intelligence cycle different from that normally used by intelligence agencies. Instead of *collection*, *analysis*, *dissemination* and *prioritisation*, this more operationally-optimised cycle would consist of *finding* an elusive enemy, *fixing* him, *finishing* him, *exploiting* his inability to protect his secrets after death, and finally *analysing* these secrets. The intelligence yield from each special operation would be the driver for launching the next one.[14]

---

[12] Strobl, Staci and Lindsay, Jon R, 'Lost in Transition: Khobar Towers and the Ambiguities of Terrorism in the 1990s', in Haberfeld, MR and Hassell, Agostino von eds., *A New Understanding of Terrorism: Case Studies, Trajectories and Lessons Learned* (New York: Springer, 2009), p. 301.

[13] In Iraq, the US military prioritised hitting middle-rank managers, knowing that these constituted the key link in a terrorist network. Neutralising them was like killing the senior non-commissioned officers of an enemy army. It would break the link between the top terrorist leadership and the footsoldiers. To identify them, the US organised all information into a common Joint Special Operations Command (JSOC) cyber platform or 'intranet', which was accessible to every analyst with the requisite security clearance. The same data could be used to plan operations, and to conduct long-term trend analysis. The US also set up tactical human intelligence teams (THTs) of six-seven soldiers who operated across Iraq recruiting informers, operating with the protection of nearby US units and going on patrol with them. By 2004 there were 140 such THTs in Iraq. See Urban, Mark, *Task Force Black: The Explosive True Story of the SAS and the Secret War in Iraq* (London: Abacus, 2011), pp. 50-52 and p. 80.

[14] Crawford, George A, *Manhunting: Counter-Network Organization for Irregular Warfare* (Hurlburt Field, Florida: Joint Special Operations University, 2009), p. 10.

Owing to the extraordinarily high risks involved in this intelligence cycle (abbreviated to F3EA), the only people who can actually gather the initial data are special operations personnel. Thus, militaries faced with counter-insurgency and counterterrorism duties raise dedicated units for secretly gathering information in hostile territory. Even if this territory is under the physical control of the military, it still needs to be treated as hostile if the sympathies of the local population lie with the insurgents/terrorists. The British learnt this in Northern Ireland, when their efforts to penetrate the IRA ran into difficulties due to high levels of local hostility.[15] After recognising that traditional means of acquiring intelligence through police informers and MI5 agents would not suffice (because the IRA had a good idea of who these sources were), the British Army set up two specialist units: The 14th Intelligence Company and the Force Research Unit (respectively abbreviated to 14 Int and FRU). One dealt with clandestine surveillance of terrorist hideouts, while the other recruited and ran deep-penetration agents.

Despite the sophisticated training of their operatives (discussed in chapter 4), both 14 Int and FRU still faced considerable difficulty in monitoring the IRA. Only about 5 per cent of all people 'pitched' by FRU handlers eventually agreed to become informers, and even then, they were not IRA insiders, but merely people who lived in localities where the IRA had a strong presence. For highly sensitive information, FRU depended upon defectors-in-place: IRA turncoats who themselves made contact with the British military and let it be known that they were willing to provide information. Usually, their motives were straightforward cases of seeking revenge against either the IRA organisation as a whole, or specific operatives within the organisation with whom they had quarrelled.[16]

---

[15] Rennie, James, *The Operators: The Army's Top Secret Elite* (London: Century, 1996), p. 136 and p. 177.

[16] Ingram, Martin and Harkin, Greg, *Stakeknife: Britain's Secret Agents in Ireland* (Dublin: O'Brien Press, 2004), pp. 36-44 and p. 56.

Besides the intrinsic difficulty of acquiring information, special operations personnel face another problem: Getting others to believe it. Often, the data obtained from physical observation in hostile territory, or from document seizures and prisoner interrogation, is disregarded at the level of higher headquarters if it cannot be corroborated. Yet, lack of corroboration does not in itself mean that the information is false. In some cases, special operations units have to go to absurd lengths just to prove that their reports from the field are accurate. During the Vietnam War, an intelligence analyst questioned a SOG-MACV reconnaissance report that the North Vietnamese were using elephants to transport supplies along the Ho Chi Minh Trail, citing lack of evidence. The answer came back a few weeks later, when an SOG team leader returned from a reconnaissance mission on the Trail with a plastic bag full of elephant dung, which he placed on the debriefing room table.[17]

Despite these drawbacks, special operations are a vital adjunct to intelligence at both strategic and tactical levels. Evidence gathered from these operations can expose a foreign covert operation, sometimes even to the extent of forcing its closure. For instance, in July 1987, naval mines mysteriously began appearing in the Persian Gulf, threatening both merchant shipping as well as US and other naval vessels. Though suspecting that the mines had been laid by Iran, Washington had no proof. Iranian officials denied involvement, going to the extent of claiming that the placement of mines was due to the 'Hand of God'. Their denials turned to embarrassment in September, when US Navy SEALs boarded an Iranian vessel which had been secretly tracked by aerial surveillance. Unaware that they were being observed, the Iranian sailors were dropping mines into the sea even as American helicopters strafed them, killing several and forcing the rest to abandon ship. The subsequent boarding of

[17] Plaster, *SOG*, p. 43.

the vessel yielded a proverbial treasure trove of intelligence on a secret Iranian plan, code named 'Ghadir', to mine the Persian Gulf. So damning was the evidence, together with the testimony of the prisoners, that Iran immediately halted all mine-laying operations.[18]

Domestic Versus Foreign Operations

In a postmodern age where governments have de-emphasised nationalist and hyper-patriotic narratives, it is difficult to find volunteers for the dangerous work of spying. Traditionally there have been four motives for becoming a spy: Money, ideology, compromise and ego (MICE). With the end of the Cold War, ideology has lost much of its appeal in postcolonial societies that are driven by newly-unleashed materialistic urges. However, terrorists are still fired by ideological zeal, irrespective of whether the fuel that drives them is left-wing or right-wing ideology.[19]

Instead of expecting intelligence agencies to recruit well-placed 'moles' inside terrorist organisations therefore, SOF need to recognise that they must develop their own sources. The best results in counterterrorism are obtained when intelligence is produced at the same level at which it is consumed.[20] That means that planning for a special operation must be supported by strong in-house capabilities for gathering intelligence at the unit level. The US Army Special Forces (the Green Berets) have long understood this principle. Each 12-man Green Beret team has two operations and two intelligence non-commissioned officers. Each battalion has three signals intelligence teams that can handle tactical interception of enemy communications, as well as electronic warfare tasks.

---

[18] Crist, *The Twilight War*, pp. 297-299.

[19] Hughes, Geraint and Tripodi, Christian, 'Anatomy of a Surrogate: historical precedents and implications for contemporary counter-insurgency and counter-terrorism', *Small Wars & Insurgencies*, Vol. 20, No. 1 (2009), p. 10.

[20] Kovacs, Amos, 'Using Intelligence', *Intelligence and National Security*, Vol. 12, No. 4 (1997), p. 162.

Limited capabilities for prisoner interrogation (which require language skills) are dispersed throughout the ranks in order to ensure quick turnaround on raw data obtained after a firefight.

There is a basic rule of thumb: The smaller a special operations team, the more intelligence it requires.[21] When operating in hostile territory, such teams must necessarily be small, in order to reduce logistical complications as well as the risk of detection during ingress and egress. This in turn means that they require even more intelligence than would be the case in friendly territory. Although specific information is hard enough to obtain when a special operation takes place in permissive environments, it becomes infinitely more difficult when the operation takes place in an enemy country.

The Selous Scouts of Rhodesia (now Zimbabwe) best illustrate this point. During the 1970s, they developed sophisticated methods to gather intelligence both domestically and on foreign soil. Domestically, they elevated the tactic of using 'pseudo-gangs' to a fine art. Such gangs consisted of former insurgents/terrorists who agreed to switch sides after being captured and put their knowledge of insurgent passwords and operating methods at the security forces' disposal. Together with serving members of the Selous Scouts, they would go into the field, masquerading as genuine insurgents. Whenever they came into contact with another group of insurgents, they would initially bluff their way to meet the senior guerrilla commander in the area, before killing or kidnapping him.[22] Since they were operating within Rhodesian territory, they could call upon reinforcements at short notice if they were discovered too early and needed help to escape from an ambush.

---

[21] Gatchel, Theodore L, 'Pseudo Operations: A Double-Edged Sword of Counterinsurgency', in Norwitz, Jeffrey H, *Armed Groups: Studies in National Security, Counterterrorism and Counterinsurgency* (Newport, RI: US Naval War College, 2008), p. 62.

[22] Hughes and Tripodi, 'Anatomy of a Surrogate', p. 5.

However, when operating across international borders, the Selous Scouts had no protection if they were discovered. They had to operate on a completely deniable basis, similar to that of professional intelligence agencies. Reconnaissance of guerrilla bases in neighbouring Mozambique and Zambia was thus carried out using South African staging posts as cover, to disguise the origins of the operatives deployed. Whenever one was caught, he would not be claimed no matter how severely he was tortured by local security forces. The Rhodesian intelligence service provided help by obtaining false documents for the reconnaissance teams, but otherwise adopted a hands-off attitude. In this context, Selous Scouts members swiftly learned to think like covert operatives instead of soldiers; they acquired a street smart ruthlessness that only comes from the desperation to survive.[23]

US Army ISA: A Model to be Emulated

Perhaps the only well-documented example of a special operations unit dedicated to gathering intelligence overseas is the US Army Intelligence Support Activity (ISA). Set up in March 1981, it was tasked to provide intelligence and logistical support to assault units from the US special operations forces. The immediate trigger for its establishment was Operation Eagle Claw—the abortive attempt by Delta Force and Army Rangers to rescue Americans being held hostage at the US embassy in Teheran. A six-man commission of enquiry found that information supplied by the CIA had been patchy and often irrelevant to the highly specific needs of the rescue force.

The commission noted that neither the CIA nor the US SOF community had a cadre of operatives who could operate autonomously in hostile territories, surreptitiously gathering information without any local support. Language skills were (and

[23] Reid-Daly, RF, *Pamwe Chete: The Legend of the Selous Scouts* (Welteverdenpark: Covos-Day, 2001), p. 193 and pp. 403-409.

still are) quite rudimentary in the US Army Special Forces, with only 28 per cent of all ranks possessing even basic competency in a foreign language.[24] The CIA's paramilitary operatives were ex-military, and had the same problem, giving them no advantage in urbanised environments, where the emphasis was on blending in with the local population. Recognising that special operations forces needed to be guided towards their targets by personnel who were already on the ground, and that in certain contexts, sympathetic locals could not be found to do this job, the Army created ISA. The unit was intended to gather 'actionable intelligence' in high-risk areas where no other military or civilian agency could operate.[25]

Besides gathering intelligence, ISA was also supposed to create agent networks for logistical purposes. The CIA tended to focus on recruiting high-level sources: Diplomats, military commanders and politicians, but none of these would be useful in putting together the nuts and bolts of a special operation. For that purpose, ISA recruited taxi drivers, hotel managers and staff working in cargo haulage companies. Although ISA and the CIA did not compete for access to the same types of agents, there was still a strong professional rivalry between the two, which hampered the unit's effectiveness. In one case, the CIA station chief in Islamabad compromised the ISA representative in Pakistan, because he wanted to thwart ISA's cross-border agent recruiting programme in Iran.[26]

To this day, the unit's operations remain a closely-guarded secret, but it is possible to provide a rough outline. In the early 1980s, ISA sent operatives to the Middle East to reconnoitre ingress routes for US hostage rescue units. In the event that any American diplomats

---

[24] Cline, Lawrence E, *Intelligence, Special Operations and Counterterrorism: Improving the Links*, Paper delivered at the International Studies Association Annual Convention, March 2004, p. 26.

[25] Church, George J, 'The Secret Army', *Time*, 31 August 1987, pp. 12-14.

[26] Crist, *The Twilight War*, pp. 62-63.

were taken captive, as had happened in Iran, the information gathered by these advance teams would be crucial in kick-starting a rescue mission. The teams photographed and videotaped the US embassies and their neighbourhoods. Information thus gathered was entered into a software programme that allowed viewers to virtually 'walk' through a room or corridor, much as would be the case in a live operation.

One of ISA's earliest operations was code named Winter Harvest and it involved the rescue of a US Army brigadier who was held hostage by Italian leftist terrorists. American and Italian officials had no idea where the captors were keeping their victim. By using Huey helicopters fitted with sophisticated listening devices, ISA eavesdropped on the handheld radios being used by the kidnappers. After narrowing down the search area to a triangle composed of three cities, the unit's analysts studied patterns of household electricity usage, sector by sector, until they finally found an apartment where consumption had shot up since the day of the kidnapping. Italian police rescued the brigadier shortly thereafter.

Signals intelligence was ISA's core strength. The CIA and the US National Security Agency did not have equipment capable of intercepting short-range tactical radio transmissions. However, in counterterrorist and counter-insurgency operations, this is precisely the level at which electronic intercepts have the greatest 'actionable' value. The ISA filled the gap by acquiring a fleet of civilian aircraft, both fixed and rotary wing, and flying them to trouble zones across the world under cover of shadow companies. The aircraft were crammed with frequency-scanning and direction-finding equipment, allowing ISA analysts to listen in on the communications networks of insurgent and criminal groups. The information thus gathered was shared with local security forces. Among the most famous victims of ISA's signal interception operations were the Colombian drug lord

Pablo Escobar and Iraqi dictator Saddam Hussein, as well as the latter's two sons Uday and Qusay.[27]

The unit's personnel were divided into three work streams: Signals intelligence, human intelligence and direct action. Thus, besides obtaining information from both technical interception and its own network of overseas-based informers, ISA was also capable of carrying out commando raids if other units such as Delta Force were not readily available. Control of the unit alternated between operations and intelligence experts from the US Army, with the commanding officer and his second-in-command (2/IC) complementing each other. When the 2/IC eventually took over as the unit CO, he would appoint a new 2/IC who had the same background as the departing commander. By this arrangement, ISA maintained a balance between intelligence and operational expertise at the command level.

The ISA was a rare type of special operations unit. Not even the British, who have several decades' experience in counter-insurgency and counterterrorism, had anything like it. In April 2005, they established two units directly modelled on ISA—the Special Reconnaissance Regiment (which dealt with agent-running) and the 18th United Kingdom Special Forces Signal Regiment (which handled all tactical interception tasks). Through creating these units, the British sought to reduce the time lag between intelligence reporting and follow-up action, by co-locating intelligence capabilities with the operational personnel who would utilise the output.[28]

Implications for India

India urgently needs a capability for gathering specific and perishable information from enemy territory, if it is to carry out special operations

[27] Skelly, Joseph Morrison, 'A Review of The Activity's Active Intelligence', *International Journal of Intelligence and CounterIntelligence*, Vol. 21, No. 2 (2008), pp. 419-425.

[28] Cline, Lawrence E, *Intelligence, Special Operations and Counterterrorism: Improving the Links*, Paper delivered at the International Studies Association Annual Convention, March 2004, p. 26.

in Pakistan. To minimise the risk of a broader conflagration from nuclear weapons, any commando raid must be supported by detailed intelligence that allows the strike force to approach the target stealthily, neutralise it in a shock assault, and get out before Pakistani security forces have recovered their wits. Even at the best of times, this will be difficult to do, perhaps insurmountably so. However, there is a silver lining: Past special operations from across the world indicate that *when the enemy relies on secrecy rather than physical barriers as his main source of protection, he becomes vulnerable to tactical surprise.* Complacent in the thought that no one can find him, he presents a lesser challenge than if he is in a fortified location.

The rescue of the Italian dictator Benito Mussolini by German commandos is a prime example. Military historians have tended to focus more on the Germans' novel use of gliders to approach Mussolini's mountain-top prison. However, the actual rescue mission was a relatively simple affair when compared to the intelligence problem of locating Mussolini in the first place. His captors knew that the Germans were looking for him and would mount a rescue if they discovered his whereabouts. So they kept moving him around, in some cases causing the Germans to just miss him by a few hours. Eventually a stray signal intercepted by German intelligence revealed that a mountain in central Italy had been sealed off by dissident forces. Through cross-checking with other scraps of information, the Germans deduced that Mussolini was being held at this site. Had it not been for that one slip in communications security on the part of Mussolini's enemies, the Germans might not have got to him before the Allied Powers did. Over-dependence on secrecy made the Italian dissidents susceptible to surprise when the German gliders swooped in, since they had not prepared for such a contingency.[29]

While planning special operations against Pakistan therefore, India should concentrate on hitting targets that are hard to find,

---

[29] Foley, Charles, *Commando Extraordinary* (London: Cassell, 1998), pp. 64-65.

precisely because these targets once attacked, will be hard to defend. Besides extremely detailed intelligence, which has to be updated in real-time, such operations would also require innovative methods of transportation. Only then will the raiding force be able to attain tactical surprise, by approaching the enemy from a direction where he is least expecting an assault. This is a problem of improvising logistics, which is the focus of the next chapter.

3

Logistics

This chapter argues that there are two ways of arranging logistics for special operations. One is through technology—aircraft and boats that are customised for such operations. The other is through deception, primarily involving the use of disguise and local sympathisers. The former is more associated with the Anglo-Saxon model of special operations, derived from British and American experiences in World War II and post-Cold War counter-insurgency campaigns. The latter has a less illustrious pedigree—terrorist groups, repressive governments and the German Wehrmacht. Yet, it is no less useful; in a war between terrorists and governments, the side with more scruples pays dearly.

Terrorists have established support networks across the globe by tapping into diaspora communities. When Lashkar-e-Taiba (LeT) conducts reconnaissance for attacks, it seeks to outflank security screening at Indian airports by using itinerant agents such as the Pakistani-American Daood Gilani alias David Headley. In doing so, it is copying a tactic first developed by the Lebanese group Hizballah, to strike at Israeli interests across the globe. During the 1980s, using Lebanese businessmen who had migrated to South America, Hizballah created a funding network on the continent which also doubled as a logistics platform for attacks.[1] In 1992 and 1994, it was able to conduct vehicle-borne suicide bombings of the Israeli embassy and cultural centre in Buenos Aires, the capital of Argentina. The Israeli intelligence service Mossad was stunned by this development, having never expected that Hizballah was capable of launching a major attack thousands of miles from its home base.

LeT is using the Pakistani diaspora in the Middle East and in the West to prepare for long-distance 'indirect' attacks on India. These attacks are meant to unfold through several layers of cut-outs and intermediaries, which will help mask their connection to the LeT leadership and thereby, to the ISI. India cannot use its diaspora to carry out retaliatory strikes on ISI officials or LeT members in third countries—it has to confine its reprisals to Pakistan itself. Yet, as this chapter shall argue, it is not well-prepared to use technology in logistical support of these reprisals. Instead, India has to use the terrorists' own methods against them, and rely on deception to get in close proximity to targets.

Special Operations and Air Power

After World War II (WWII), American and British special operations forces became accustomed to operating under conditions of air superiority whilst on counter-insurgency operations. They did not even have to confront the threat of man-portable air defence systems (MANPADS) that the Soviets faced in Afghanistan. Thus, the Anglo-American model of SOF usage in countering irregular warfare emphasised rapid air mobility and close air support. The use

[1] Constanza, 'Hizballah and Its Mission in Latin America', pp. 199-200.

of local partners was not given as much attention as it had been during the World War.[2] Admittedly, the US Army did raise Special Forces Groups (Green Berets) to organise guerrilla operations behind enemy lines in the event of another major global war, but these were pro-insurgency campaigns, not counter-insurgency. Furthermore, wherever these units were deployed, they could still count on an aerial shield from the US Air Force.[3]

For example, air power was crucial to the success of SOG-MACV special operations in Laos and Cambodia, even though these countries were technically neutral in the Vietnam War. SOG personnel were particularly fond of calling in ARCLIGHT missions against North Vietnamese convoys plying the Laotian part of the Ho Chi Minh Trail. Each mission consisted of up to twenty-four B-52 bombers dropping as many as 2616 five hundred pound bombs over a twelve-square mile area, annihilating everything unlucky enough to be there.[4] In Cambodia, aircraft were used mainly for transportation purposes because the US hoped to win over the Cambodian government from North Vietnam's influence, and therefore limited the damage that SOG personnel could cause within the country.

This dependence on air power was an obvious strength of US special operations, and an obvious weakness. North Vietnamese studies of SOG-MACV methods soon identified the time and place of ingress as being the moment and location of greatest vulnerability. Accordingly, Landing Zone (LZ) Watcher teams were set up (as already mentioned in Chapter 1) to monitor possible ingress points and report any effort by the Americans to activate them. Whenever a

[2] Lillerud, Boye, 'Flipping the COIN: Unity of Effort and Special Operations Forces', *NUPI Working Paper No. 752*, p. 5, http://english.nupi.no/content/download/8136/81940/version/2/file/WP-752-SIP-16-Lillerud.pdf, 9 October 2013.

[3] Biddle, Stephen D, 'Allies, Airpower, and Modern Warfare: The Afghan Model in Afghanistan and Iraq', *International Security*, Vol. 30, No. 3 (2005), pp. 162-167.

[4] Plaster, *SOG*, p. 50

US reconnaissance team was observed at one of these sites, the favoured North Vietnamese tactic was to stealthily track the team as it left the LZ until the helicopters that had brought it were safely back across the international border. The border was marked by a high mountain range which would severely curtail radio communications between the SOG team and the helicopters, thus ensuring that the team could not be extracted when the North Vietnamese finally launched their ambush.

Even without the threat of MANPADS, American aircraft over Vietnam faced a grave threat from ground fire. The toll inflicted in terms of airframes might have expanded the already widespread view in the US government that Vietnam was an unwinnable war. Over the course of eleven years, the US Air Force lost 1737 fixed-wing aircraft to enemy fire.[5] The Army's losses were worse—almost 3600 helicopters were shot down by Viet Cong insurgents or North Vietnamese troops.[6] More than the fixed-wing aircraft, helicopters were crucial to US special operations. Each helicopter was to an American special operations soldier what the horse was to a Boer commando in the year 1900—a means of mobility, a carrier of supplies and a conveyor of firepower.

Throughout the twentieth century, control of the skies was a prerequisite to successful raids by the US SOF, and its loss was the cause of catastrophic failures, such as Operation Irene in 1993—the Battle of Mogadishu. All told, the raid on Somali warlord Mohammad Aideed's stronghold involved nineteen helicopters and one surveillance aircraft, with the ground component of the strike

---

[5] Grant, Rebecca, 'The Crucible of Vietnam', *Air Force Magazine*, dated 2 February 2013, http://www.airforcemag.com/MagazineArchive/Pages/2013/February%202013/0213vietnam.aspx 9 October 2013.

[6] Kapila, Viney, *The Indian Air Force: A Balanced Strategic and Tactical Appreciation* (New Delhi: Ocean Books, 2002), p. 50. This same source also asserts however, that 2561 US fixed-wing aircraft were lost in action in Vietnam.

force initially being limited to just twelve vehicles.[7] When Somali militiamen shot down one helicopter, US forces activated their contingency plan to rescue downed aircrew. However, when a second helicopter was shot out of the sky, the Americans found themselves over-stretched by the need to rescue two sets of crew members, separated from each other by several urban blocks teeming with hostile gunmen.[8] What began as a textbook commando mission degenerated into a prolonged series of streetfights, with the Americans having to focus on survival despite having attained their initial objective. This psychological and tactical shift happened largely because the Somalis understood how crucial air power was to the self-confidence and tactical ability of the US SOF, and managed to strike at it.

RESOURCE SCARCITY

When a country faces a shortage of aircraft, or a lack of adequately trained pilots, even the best special operations soldiers are rendered little more useful than elite infantry. The Russians discovered this in Chechnya, when their heli-lift capability was degraded by 55 per cent over three years (1999-2002), due to attrition by Chechen MANPADS. There were no funds to obtain replacement aircraft—a single Spetsnaz (special operations forces) detachment in Afghanistan had more air support in the 1980s than all Spetsnaz forces in Chechnya during the early 2000s. This abysmal state of affairs was brought about by the collapse of the Soviet Union in 1991 and the economic crisis that followed. Not only were aircraft scarce, but virtually all well-trained and highly-experienced pilots left the military, fed up with the low pay. For those who stayed, average annual flying time fell by 90 per cent, as part of an economising trend. Helicopter pilots, thus, lacked the basic low-altitude combat

---

[7] McKinney, Mike and Ryan, Mike, *Chariots of the Damned: Airborne Rescues from Vietnam to Kosovo* (London: HarperCollins, 2001), pp. 147-149.

[8] Bowden, *Black Hawk Down*, p. 491.

flying skills in Chechnya which their predecessors had learnt through much hardship in Afghanistan. Without tactical airlift and close air support capability, Spetsnaz troops in Chechnya faced tremendous difficulties in swiftly responding to guerrilla ambushes of regular infantry troops.[9]

Even the US, which of all countries never seems short of cash for its armed forces, discovered that merely having a large number of airframes is not enough; these need to perfectly match the requirements of a particular special operation. Such matching is difficult to achieve without a dedicated special operations aviation wing. During Operation Eagle Claw (the abortive rescue of American hostages in Iran in 1980), the US used helicopters that were optimised for maritime operations, not desert flying. Maintenance crews forgot to insert sand filters into the engine intakes. The result was that when the helicopters were hit by an unexpected sandstorm while flying over Iran, the intakes became clogged. Airflow to the engines was reduced, forcing the helicopters to fly slowly and use up more fuel. Meanwhile, the pilots struggled to cope with the challenge of flying blindly in close formation while having to maintain radio silence. They had little experience of night flying, and whatever they did have, had been gained under optimal training conditions, not inclement weather.[10]

Once the US rescue mission ended in a debacle, after-action reviews established that logistics had been correctly identified as the weak point in the attack plan, even while it was being drawn up. Eager to push through an ambitious idea, the Joint Chiefs of Staff and the National Security Council had nevertheless railroaded the plan through to the White House, where it was endorsed by a

[9] Kramer, 'Guerrilla Warfare, Counterinsurgency and Terrorism in the North Caucasus', pp.223-235.

[10] McKinney and Ryan, *Chariots of the Damned*, pp. 51-77.

president keen to look tough in an election year.[11] However, lack of objectivity was a secondary or even tertiary reason for the mission failure in Iran: The more immediate cause was lack of dedicated aircraft and aircrew for special operations. To address this deficiency, the US Army raised the 160th Special Operations Aviation Regiment (160th SOAR), thus creating a new kind of helicopter pilot. Instead of the methodical mindset of regular fliers, special operations aviators needed to be quick-thinking daredevils with extremely fast reflexes. Such reflexes could be enhanced through training, but they could not be created from scratch if the potential for them did not already exist in a pilot.

PILOT TRAINING AND SPECIALISED AIRCRAFT

To support special operations, the US has established a mini-air force that is concentrated in the Special Operations Command (SOCOM) and its ultra-secret sub-component, the Joint Special Operations Command (JSOC). Consisting of both Army and Air Force personnel, the aviation element within SOCOM trains intensively for high-risk missions, using a variety of aircraft. The pilots are psychologically conditioned to carry out near-impossible manoeuvres with their airframes, since this can often make the difference between a successful mission and a failure, or even no mission at all. For an illustration of this point, one only needs to look to the Son Tay raid of 1970.

SON TAY: THE PROTOTYPE

Son Tay was a prisoner-of-war (POW) camp in North Vietnam. During the summer of 1970, US intelligence agencies became convinced that a number of Americans were being held there. A rescue operation was planned for November, using fifty-six men

---

[11] Smith, Steve, 'Groupthink and the Hostage Rescue Mission', *British Journal of Political Science*, Vol. 15, No. 1 (1985), pp. 120-122.

from the US Army Special Forces as assaulters. Helicopter pilots were drawn from the Aerospace Rescue and Recovery Service—as rescuers of downed bomber pilots, they were the only branch of the US military that had experience of deep-penetration helicopter sorties over North Vietnam. Prior to the raid, the Aerospace Rescue and Recovery Squadron (ARRS) pilots conducted a total of 280 rehearsal sorties totalling 790 flight hours. On the night itself, they were escorted by fixed wing aircraft, which were meant to provide protection against hostile fighters. From the east, a US carrier group situated in the Pacific Ocean launched air strikes against the North Vietnamese coastline, to distract radar operators from the rescue force, which entered the country from the West.

Owing to fears for the POWs' safety once camp guards realised that a rescue mission was underway, it had been decided to crash-land a helicopter directly into the cramped prison compound itself. This was extremely risky, since the type of helicopter used was quite large because of the flight distance that needed to be covered. After several rehearsals, the pilots had grown confident of making a controlled night crash into an extremely tight space between the camp buildings, even while coming under simulated fire. Their flying skill was crucial to allowing the 14-man Green Beret team assigned to the helicopter to storm out even before the guards were aware of what was happening.[12] The raid occurred on 21 November 1970. Tactically, it was superbly executed and became a model for subsequent US special operations. However, its success was negated by the discovery that intelligence on the camp had been outdated—the POWs had been relocated to another area some months previously.

The Son Tay raid, with its reliance on technology and extensive rehearsals, serves as a good example of how the US conducts special operations. It was completely dependent on air power for both transport and fire support. It was planned on the basis of intelligence

[12]McKinney and Ryan, *Chariots of the Damned*, pp. 30-37.

obtained from aerial and satellite reconnaissance, not human intelligence. It involved the deliberate sacrifice of an aircraft in order to enhance the assault force's chances of gaining surprise. It involved excellent combat flying skills, honed to perfection through costly and time-consuming dry-runs. Finally, it was carried out at an optimal time—during a typhoon, which impeded the North Vietnamese response.[13] The discerning reader would be able to notice numerous similarities between Son Tay and the 2011 Abbottabad raid. In the latter too, air power and geospatial intelligence were crucial enablers, a helicopter was crash-landed (though out of necessity rather than pre-planned choice), pilot skills saved the assaulters from certain death in the crash, and the timing of the raid was optimal—an exceptionally dark night, where the attackers' nightscopes gave them a decisive advantage over the defenders.[14]

ABBOTTABAD: THE CULMINATION

On the Abbottabad raid, the pilots used nap-of-the-earth (NOE) flying techniques, staying close to the ground while travelling at high speed. That, combined with the special stealth features of their modified MH-60 helicopters (each reportedly costing USD 60 mn), ensured that they were able to approach the target location undetected by Pakistani air defences. Admittedly, this involved a slight risk: NOE tends to induce airsickness, potentially detracting from the combat efficiency of a raiding force that has just been through a long bout of it. The operational planners sought to minimise the risk by selecting only experienced troops for the raid. Most of the SEALs who participated had been in air crashes before, and thus were capable of operating effectively even after a very bumpy flight.

[13] Carney, John T and Schemmer, Benjamin F, *No Room for Error: The Covert Operations of America's Special Tactics Units from Iran to Afghanistan* (New York: Random House, 2003), p. 100.

[14] Owen, Mark with Maurer, Kevin, *No Easy Day* (London: Penguin, 2012), p. 190.

The pilots on the Abbottabad mission were part of an elite within the US military. Over several years, the military had found that it was easier and more cost-effective to train a special operations aviator to transition from one type of aircraft to another, rather than provide special operations training to regular pilots who were experienced in flying the particular type of aircraft needed for an operation. A programme of advanced training was, therefore, introduced in the 1980s, to identify experienced aviators and train them for at least eight months in new aircraft and flying techniques. These pilots formed the pool of aircrew who would be called to provide logistics and firepower support to US special operations worldwide. Besides transport helicopters, they also could fly gunships designed for urban operations. One of these was the McDonnell Douglas 530, popularly known as 'Little Bird'. With specially designed rotor blades, it was extremely manoeuverable while also being silent. It was built for night flying, and came in two versions: A command and control (C2) model with forward-looking infrared radar and video recording system, and an attack model with a 7.62 mm minigun (capable of firing 3,000 rounds per minute) and a 2.75 inch rocket pod. The 'Little Birds' usually trained and operated in groups of three: A command model, flanked by two attack models.

It was an attack by 'Little Birds' that halted Iran's secret mining of the Persian Gulf in September 1987 (an incident already referred to in Chapter 2). With their night vision technology and low engine noise, the aircraft were able to surreptitiously track an Iranian ship during the hours of darkness as it laid mines along shipping routes. Unaware that they were being observed, the Iranians did not bother to hide what they were doing, thus providing the Americans with clinching visual evidence that Iran was behind the mining campaign. The first that the ship's crew knew of the American presence in the sky was when the helicopters, having gathered the necessary evidence and having received attack orders, swooped in with miniguns blazing. Several crew members died and the rest abandoned ship.

Infiltration From the Sea

So far, this chapter has described how air power can help special operations forces enter a hostile area and carry out reconnaissance or direct action missions. The same broadly applies to marine-based logistical platforms. Boats theoretically offer more flexibility than aircraft, since they can drop off a raiding force anywhere along an unguarded coastline and let the force approach its target overland. However, in practice marine ingress also has limitations: Few critical targets are located close to the sea, precisely because they would be vulnerable to an amphibious attack. In Vietnam, MACV-SOG's maritime operations wing, code-named OP-37, had far more success while attacking targets in friendly South Vietnam than in North Vietnam. The North was technically still a sovereign country, and US commandos could not be openly seen attacking its coastline. For this reason, SOG operations against coastal targets in the North were stand-alone attacks that were not reinforced by the US Navy if they ran into trouble. The actual raiders who went ashore were South Vietnamese commandos trained by SOG; so worried was the American government about getting caught in these covert activities that it refused American servicemen permission to directly take part in the raids.[15]

No such restraints can be seen when it comes to Pakistani covert operations against India. The Inter-Services Intelligence (ISI) had no problems about using Pakistani nationals to kill Indian civilians and foreign tourists in Mumbai, having anyway planned to deny the killers' nationality. The Pakistani army had, after all, done the same with its own soldiers in Kargil. Using LeT *fidayeen* to land directly

---

[15] Schultz, Jr, *The Secret War Against Hanoi*, p. 200.

on the Mumbai coastline was, thus, a natural choice. In retaliation, the Indian Navy's Marine Commandos (MARCOS) could have been authorised to hit at Karachi. Instead of a hijacked fishing trawler they could have infiltrated via submarine and killed Pakistani servicemen at PNS Iqbal, one of the training areas used by the *fidayeen*. An operation of this sort might still be required in the future if Pakistan continues to use terrorism as an instrument of foreign policy. India cannot under any circumstances allow nuclear weapon-enabled terrorism to be a one-sided affair. However, only southern Pakistan (primarily Karachi) lies within range of the Indian Navy, while the bulk of LeT recruitment and training occurs in the northern parts of the country. Hitting these locations requires using deception, and not technology, as a means of infiltration.

Deception as an Ingress Method

Although scholars of Western antiquity might like to reference Trojan Horse, in modern times it was the Wehrmacht that provided the first example of successful special operations, carried out primarily through deception. Germany's armed forces in 1939 had a number of immigrants from other European countries—ethnic Germans who were eager to unite their former homelands with the Reich. Towards this, they were prepared to employ their linguistic abilities and knowledge of local cultures, to carry out deep-penetration surgical strikes against critical targets.[16] While approaching these targets prior to the actual start of hostilities, they would travel in disguise and use forged identity papers. The Wehrmacht's Blitzkrieg doctrine hinged crucially on the success of these undercover troops—if they could neutralise enemy defences at the decisive moment, German Panzer formations and mechanised infantry would break through and thrust deep into the enemy's hinterland.

---

[16] Lefevre, Eric, *Brandenburg Division: Commandos of the Reich* (Paris: Histoire & Collections, 2000), p. 10.

Otto Skorzeny was a German officer who showed a particular flair for special operations, becoming famous as the man who rescued Italian dictator Benito Mussolini from a mountaintop prison in 1943. Skorzeny liked to compare direct action missions with 'carving out the brains' of the enemy—attacking in such a manner, that the enemy would be paralysed even before he knew what was happening.[17] To achieve this shock effect, Skorzeny relied on his own ingenuity, and the initiative of his troops. The goal was always straightforward: The enemy must be attacked only at that place and at that moment when he believes that he is most secure. However, precise details of the attack would have to be left to commanders on the ground who would respond to each problem as they saw fit. Often, this required tactical/technical improvisation to maintain stealth while initially approaching a target, and then using deception to further confuse defenders during the actual assault.

For example, during the Mussolini rescue, Skorzeny used gliders to transport a force of roughly 100 commandos directly to the forecourt of Mussolini's prison. Since gliders had never before been used at such high altitude, the defenders were unsure of what they were seeing. They did not know if it was a rescue effort by the Germans or just a training exercise by the Italian military. Their confusion was further compounded when an Italian general, who was pro-German, emerged from one of the gliders and demanded that they put up no resistance. Obtaining the cooperation of this general had been a masterstroke by Skorzeny's second-in-command; the Italian guards of Mussolini were unwilling to fire upon one of their own officers, and promptly made way for the German force.

Later, Skorzeny made extensive use of disguises while planning special operations against Allied armies in Western Europe. The Wacht am Rhein offensive of 1944—a push intended to stall the Allied advance towards Germany, even became notorious for allegedly

---

[17] Foley, *Commando Extraordinary*, p. 49.

breaching the laws of war. During this offensive, German commandos were outfitted with Allied uniforms, vehicles and weaponry, and tasked to sow confusion behind enemy lines. The scheme might have done considerable material damage, had the Germans' overall military situation not been so weak by that time in the war. Being the losing side, Germany did not have sufficient quantities of Allied firearms to issue the commandos, nor did it have enough fluent English-speakers. The very few Wehrmacht personnel who spoke English to a native standard, were able to move back and forth across the front, gathering intelligence on Allied troop movements. However, the majority were hamstrung by inadequate training and complete lack of operational support—there simply were not enough German aircraft to resupply the lightly armed commando detachments and keep them fighting. After the war, Skorzeny was tried for having ordered his men to fight whilst in enemy uniform, but was let off once it became known that the British had used identical methods against the Germans. His trial set a precedent: The legitimacy of a tactic must be gauged with reference to its context, not to some archaic law drawn up in a completely different threat scenario.[18] India would do well to absorb this lesson as it seeks to combat Pakistan's proxy war: Fighting fire with fire is not immoral, if the other side was the first to light a match.

OPERATION ELAND: THE MOST DEVASTATING CROSS-BORDER RAID

To fully appreciate the spectacular results that might be obtained from using deception and disguise to access a target, it is necessary to look in detail at one particular operation. The Selous Scouts was a unit of the Rhodesian Army (now the Zimbabwean Army) that specialised in pseudo-terrorist activities during the 1970s. Members of

---

[18] Weingartner, James J, 'Otto Skorzeny and the Laws of War', *Journal of Military History*, Vol. 55, No. 2 (1991), pp. 209-221.

the unit disguised themselves as tribal guerrillas and gathered intelligence by interacting with villagers who were hostile to the army. As rural unrest grew in Rhodesia and the army mounted offensive counter-insurgency operations, many of the rebels obtained shelter in neighbouring Mozambique. There, they constructed training complexes, which were co-located next to Mozambican Army units for protection.

One of these camps was a massive facility, housing approximately 5,000 trainees. Aerial intelligence and prisoner interrogation helped locate the camp, and the Selous Scouts' commanding officer lobbied for carrying out a raid. He pointed out that never before had the Rhodesian security forces got a chance to cripple the tribal insurgency with a single blow: The guerrillas were complacent, thinking they were safe in Mozambique. Their vigilance level would be low. Whatever risk there was, came mainly from the logistics of arranging an attack, as well as the political outcry that may result.

Approval to strike eventually came through from the highest level of the Rhodesian government. The Selous Scouts, having built up a formidable expertise in covert operations as a result of their pseudo-terrorist activities, were chosen as the raiding force. Their plan was to use eighty-four men in a 'flying column' consisting of four armoured cars and ten armoured personnel carriers. The vehicles were painted over to conceal Rhodesian markings. All drivers and troops who rode on the outside of these vehicles, in plain sight of the public, wore Mozambican Army uniforms. Since Mozambique was a Portuguese-speaking country, officers fluent in the language rode in the front of the column, ready to bluff their way past any security checkpoint. The training camp was located sixty miles across the international border, and the deeper the column drove into Mozambique, the lower would be the risk quotient.

The raid was timed to perfection: It was carried out when the camp security detail was recovering from a weekend of festivities following a government holiday. D-Day was Monday, 9 August 1976. H-Hour was 0810 hours, when the trainees were being mustered for morning parade. The column entered Mozambique the preceding night through dirt tracks which had been identified by cross-border smugglers working on behalf of the Selous Scouts. Telephone wires connecting towns along the infiltration route were cut as the convoy raced deeper into enemy territory. After leaving a twelve-man blocking force at a bottleneck position to hold off any reinforcements, the column entered the camp. There were fifty Mozambican Army sentries at the site itself, but only three were armed. They let the vehicles in through the main gate after a Rhodesian officer lambasted them in Portuguese.

Within a few short minutes, over a thousand insurgents were lying dead on the camp parade ground. The Rhodesian vehicles had been fitted with heavy machine guns and 20mm cannon specifically for the occasion. These weapon platforms, together with the assault rifles carried by individual soldiers, inflicted the worst single loss of manpower that the guerrillas ever suffered during the entire course of Rhodesia's tribal insurgency. One insurgent leader later admitted that the attack had been devastating to their morale and resources.[19] However, there was a political price to be paid: Once they had recovered from the initial shock, the guerrillas successfully portrayed the complex in international human rights forums as a refugee camp, engaged in humanitarian activities. All evidence of military infrastructure was removed and the Rhodesians were accused of carrying out a cold-blooded massacre of displaced persons. The Rhodesian government demanded that the exact nature of the camp be investigated by an independent commission of enquiry, but could not find many takers for this idea.[20]

---

[19] 'Rhodesia – The Nyadzonya Raid', http://www.youtube.com/watch?v=Vs7V_IBQcDg, 8 November 2013.

[20] Reid-Daly, *Pamwe Chete*, pp. 202-247.

A DOUBLE OUTLIER

The August 1976 raid is a fascinating case study for two reasons. First, at a stroke it made the Selous Scouts one of the most successful counter-insurgency units in modern history. Through the Rhodesian insurgency (known today as Bush War), the Scouts accounted for 68 per cent of all guerrillas killed. The unit's contribution was either direct (raids on training camps and supply bases inside and outside Rhodesia) or indirect (gathering intelligence through pseudo-terrorist techniques). The vast majority of its kills were scored outside Rhodesian borders, in the insurgents' safe havens, where a false sense of security made the rebels easy targets. The 1976 raid was the most devastating attack of all in terms of the destruction caused, partly because it was the first such action of the war.

Second, the raid was distinct by its non-use of air power. Originally, the Rhodesians had planned to disclaim all responsibility, and so avoided committing any aircraft in support of the ground operation. The Selous Scouts column had to rely on the soundness of its commanders' tactical judgments, as well as the accuracy of the intelligence briefings that had been provided to it, to survive. Once launched, it was an entirely self-sufficient force, using guile and daring to get to its objective and back safely. Only after the Rhodesian Army's involvement was inadvertently exposed (by a road accident that forced the abandonment of an armoured car inside Mozambican territory), did the government decide to admit responsibility. Thereafter, it started regularly carrying out cross-border raids, using aircraft to effect 'vertical envelopment' of insurgent training camps. Yet, despite the additional firepower and logistic support that subsequent special operations had, none came close to generating the same psychological impact as the August 1976 raid. Instead, as Rhodesian attacks became more frequent, the insurgents learnt to absorb the damage caused and the Mozambican Army, from being a helpless observer, started taking proactive measures to intercept the raiders. Over time, this meant that the strategic effectiveness of cross-border raids by the Rhodesians diminished significantly.[21]

INDIA SHOULD RELY ON DECEPTION-BASED SPECIAL OPERATIONS

There is a similarity between the operational challenges facing India today, and those that confronted American and Rhodesian commanders in the 1960s and 1970s. In all three cases, security forces fought an irregular adversary who enjoyed safe havens across an international border. For political reasons, none of the defending countries could openly carry out large-scale military attacks on these sanctuaries. However, relying on technology-based special operations alone had its own problems.

North Vietnam had one of the densest and most sophisticated air defence systems in the world, requiring that any US raid should be provided with massive aerial cover. This, in turn, demanded that the choice of target be carefully selected to appear as 'legitimate', since the origins of the raid could not be plausibly denied. Rescuing POWs from North Vietnamese captivity was one such objective, but otherwise, there was little else to publicly justify an attack. For much of the war, the US confined its special operations to countries like Laos and Cambodia, where US air supremacy was unopposed. This allowed the operations to be conducted with little risk of exposure and international condemnation. Actual commando strikes against North Vietnam itself were kept to a minimum.

The Rhodesians were worse off than the Americans, since their country (which was at the time, ruled by an unpopular minority government) was under international sanctions. Buying new aircraft and spare parts needed for the upkeep of existing ones was difficult. Consequently, the Rhodesians initially began their cross-border attacks by relying heavily on deception rather than technology as a

[21] Cilliers, JK, *Counter-Insurgency in Rhodesia* (London: Croom Helm, 1985), pp. 178-179.

tool of infiltration. Over time, however, operational momentum and a lack of strategic foresight led to them settling into a predictable pattern, which the insurgents adjusted to. The increased use of air power was offset by the political costs of violating a neighbouring country's sovereignty, which internationalised Bush War even further.

India now faces a stark choice: It can build up its technology-based delivery platforms to carry out cross-border raids, or invest in human agents who will facilitate genuinely 'deniable' operations. The US has tended to prefer a technology-based approach to special operations because this plays to the Americans' traditional strength. Building agent networks is time-consuming and highly risky for the actual field operatives who do the groundwork. Moreover, as the Americans discovered in Iran, it is easier to recruit somebody to act as a spotter, logistician or courier, than as a saboteur. No matter how much a person might hate his or her government, they will usually hesitate to commit an act of violence themselves. This requires that foreign nationals carry out the actual attack, which can have hugely negative repercussions if they are caught and exposed as working for the US government.[22]

In our own context, the terrain along the Line of Control (LoC) favours a quick heli-borne raid against LeT training camps. However, the flying skills that nap-of-the-earth flights require would be difficult to develop in secret. India faces a serious problem in maintaining operational secrecy, as American and Chinese spy satellites, not to mention dozens of foreign intelligence agencies active in Delhi, are likely to pick up telltale signs of a special operation that uses aviation assets. To truly catch Pakistan off guard, it is also necessary to deceive the West, China and the Arab world about our immediate intentions. This would anyway be next-to-impossible in the aftermath of a major

[22] Crist, *The Twilight War*, p. 80.

terrorist strike on India, when expectations of a retaliatory attack on Pakistan would be high. Furthermore, given how dependent an aerial assault would be on prevailing weather conditions, it is a safe assumption that the Pakistanis would be able to second-guess exactly when the Indian Army might launch a heli-borne strike. Meteorological forecasts, after all, can be produced on both sides.

The solution would, therefore, be to build agent networks on an ongoing basis, to be activated upon orders from the top political leadership. Indian Air Force (IAF) pilots currently lack the flight training to carry out highly precise manoeuvres that American pilots from SOCOM aviation units can. There is no culture of joint operations planning in Indian armed forces. Even if current proposals to create a tri-services special operations command were to gain momentum, their implementation would probably be hampered by bureaucratic infighting between the various armed forces, as well as with the Ministry of Defence (MoD). Lastly, there is no massive asymmetry of air power capabilities between India and Pakistan (at least none that concerns a short-duration engagement).[23]

For the Abbottabad raid, the US had deployed two aircraft carriers in the Arabian Sea—the USS Carl Vinson and the USS Enterprise. In addition to the two MH-60 helicopters which carried the actual assault troops, there were two CH-47 Chinooks carrying reinforcements, should any be needed on the ground. Up to seven fixed wing aircraft were operating in the Federally Administered Tribal Areas (FATA) on the night of the raid, with permission of Pakistani authorities, ostensibly on counter-insurgency duties against the Taliban. These included EC-130E/H electronic warfare aircraft, an MC-130 command and control aircraft, an AC-130 gunship, two Airborne Warning and Control Systems (AWACS) and two F-15C

[23] Sood, VK and Sawhney, Pravin, *Operation Parakram: The War Unfinished* (New Delhi: Sage, 2003), pp. 156-165.

fighters on combat air patrol. Had the Pakistanis moved to scramble their fighters, all of these aviation assets would have been pressed into service to protect the withdrawing SEAL team.[24]

LACK OF EXPERIENCE IS A HANDICAP

Air power greatly simplifies the execution of a special operation, by reducing (but not eliminating) the need for extensive on-ground contacts amongst a sympathetic population.[25] Yet, India has little or no experience of carrying out special heli-borne operations (SHBOs) in hostile territory. The one time such an operation was attempted on a large scale—the Jaffna University heli-drop of 1987—it turned into a disaster. Inadequate preparation time, too few aircrafts, weak communications security and incompetent staff work gave the enemy (Tamil Tigers) a perfect opportunity to ambush the raiding force even before the operation had fully commenced.[26]

While past failures should not deter us from future efforts, we need to appreciate the challenges of carrying out technology-based special operations against an equally competent adversary such as Pakistan. Given how much we have suffered from Pakistani covert assaults over the past three decades, there is a strong case for applying the lessons learnt, in reverse. Our wealth of experience in how to combat irregular warfare through low-tech means should be utilised to wage irregular warfare against Pakistan, with special operations forces serving as the cutting edge.

---

[24] Subramaniam, Arjun, 'Air Power in Stability and Anti-Terrorist Operations', *Air Power*, Vol. 6, No. 3 (2011), pp. 49-50.

[25] Datta, Saikat, 'Air Power and Special Operations', *Air Power*, Vol. 2, No 2 (2005), p. 99.

[26] 'Descent Into Danger – The Jaffna University Helidrop', http://www.bharat-rakshak.com/IAF/History/1987IPKF/Chapter3.html, 11 November 2013.

4

Training

This chapter makes an argument for training special operations forces primarily to conduct tradecraft-based urban surveillance, rather than endurance-based rural reconnaissance. It holds that the dominant template for SOF training will increasingly become outdated, as military robotics and unmanned aerial vehicles take over many of the functions previously carried out by SOF.[1] Moreover, the kind of missions that India will need to conduct against a nuclear-armed Pakistan cannot adhere to established patterns. Since the Pakistani army and its special forces are trained according to a Western-derived model (as we are), there is currently little that we can do to surprise them strategically.

The chapter once again differentiates between British and German models of special operations, that developed during World War II. The British preferred what might be called an 'intrinsic

[1] Fitzsimmons, Michael, 'The importance of being special: Planning for the future of US special operations forces', *Defense & Security Analysis*, Vol. 19, No. 3 (2003), p. 211.

strength/acquired skill' training paradigm. The Germans did the opposite, choosing an 'acquired strength/intrinsic skill' paradigm. Following the war's end, the British model has come to be considered as the gold standard, in part because it was copied by the Americans. Although both the Americans and the British have also made efforts to create 'skill first' units, which focus on urban intelligence-gathering and are highly regarded within their respective special operations communities, these are still secondary to direct action units, which enjoy more political visibility.

High Risks = High Costs

Training sounds like a mundane subject, but it is as crucial to the success of a special operation as appropriate tasking, accurate intelligence and adequate logistics. It needs to be shrewdly invested in, because it is always expensive when done properly. The personnel who undergo SOF training handle specialised equipment that only the most competent soldiers can be entrusted with.[2] In countries such as the US, the annual cost of training, transporting and sustaining a single SOF soldier in the field can easily exceed a million dollars. This is about the same as the cost of acquiring a cruise missile. Yet, cruise missiles cannot hunt down and surgically neutralise an ill-defined military target hiding amidst a civilian population. A well-trained special operations force can.[3]

Besides the inherent cost-effectiveness of SOF, they also deserve high-quality training because of the extremely hazardous nature of their work, when compared to other branches of the military. On average, SOF units across the world have an operational fatality rate twenty-two times higher than other units. If training deaths

[2] For instance, the four-tube enhanced peripheral night vision goggles which US Navy SEALs used on the bin Laden raid cost USD 65,000 apiece. Owen with Maurer, *No Easy Day*, p. 204.

[3] Couch, Dick, *The Warrior Elite: The Forging of SEAL Class 228* (New York: Three Rivers Press, 2001), p. 308.

are included, then the fatality rate increases to around forty times.[4] Realistic training is dangerous training, but the best units consciously choose to expose their aspiring members to such risks under relatively controlled conditions, rather than leave them to the mercies of fate in an actual combat situation. Political and military decision-makers, thus, have a responsibility to ensure that the training provided to SOF is the best that a nation can provide.

Generally, training for special operations forces proceeds according to the maxim that 'in order to get one good man, it's necessary to begin with five good men'.[5] Rapid attrition of unsuitable candidates early in the process (usually during discrete 'selection' and 'probation' phases) helps focus instructional efforts on those who are really motivated and capable of contributing to a unit. However, the parameter that decides the attrition rate can vary: In the mainstream Anglo-American model, it is physical toughness, as measured in terms of strength and endurance. In the less widely used German Wehrmacht model, it is the ability to move unnoticed in a densely populated locale, using language skills and disguises. On average, the second model produces soldiers who are smarter than those of the first model. Both, however, are optimised for different kinds of mission and operational environments. The British and Americans have attempted to harness some of the traits of the German model to produce all-rounded special operations forces. Whether they have been successful in this is another matter, as is the question of whether India should copy them.

The Intrinsic Strength/Acquired Skill Model

There seems to be a common pattern among the US and UK special operations units, in terms of training programmes and timelines. From start to finish, it takes about thirty months to fully train an

---

4 Carney and Schemmer, *No Room for Error*, p. x and p. 23.

5 Couch, *The Warrior Elite*, p. 12.

SOF soldier. The first three to four weeks are focused on selection—weeding out candidates who are physically unfit, or not very serious about joining the unit. Attrition rates are the highest at this point: Usually around eighty per cent. The high number of dropouts is primarily due to constant petty harassment from the instructors, and only secondarily due to the physical stress of the actual tests.[6] When somebody chooses to withdraw from a selection course, or is injured, he is permitted to reapply for another course, provided the instructors have developed a favourable impression of his potential. Tremendous importance is given to their judgment, for the sake of which minor rule-bending is allowed.

Those who pass selection are subjected to continuation training for roughly five months while still being formally on probation. Among other things, they learn small-team tactics, weapon drills, combat navigation and reconnaissance. Parachute training is almost always conducted at some point, not because it is of much use in actual operations, but because it is thought to be a great character-builder.[7] Escape and evasion (E&E) training is also provided, teaching candidates how to live off the land, avoid tracker dogs, and passively resist enemy interrogation (this last includes simulations of physical and verbal abuse but usually does not extend to limb-threatening torture).

After about six months, the probationary phase ends and the successful volunteers are finally considered members of the unit. They then spend the next two years improving on the basic skills they learnt previously, with a view towards identifying ways of tactical improvisation. Special operations tend to have a higher rate of success if they feature either a new transportation system (which helps in achieving surprise) or a new weapon system (which helps

[6] Ibid., p. 167.

[7] Geraghty, Tony, *Who Dares Wins: The Story of the SAS, 1950-1992* (London: Time Warner, 2002), p. 523.

in maintaining speed over target). Thus, even while deployed on active operations in Vietnam, units like SOG-MACV found time to experiment with skydiving techniques. The result was the High-Altitude/Low-Opening (HALO) method of aerial insertion. Using this, SOG reconnaissance teams could reduce their chances of being detected by radar when making night jumps over hostile territory.[8]

For the sake of realism, each unit can tailor the actual curriculum to best reflect its bureaucratic and operational reality. Delta Force, for instance, draws the bulk of its manpower from the US Army Special Forces and the Rangers. Since men in these units have already been tested on their teamwork skills, Delta instructors instead focus on studying how a candidate performs in solo tasks. The British SAS, on the other hand, puts slightly more emphasis on teamwork since a large number of its men are drawn from outside the Parachute Regiment (which is the main feeder of manpower). The US Navy SEALs lay the most emphasis of all on teamwork, viewing it as a force multiplier. This partly has to do with the unorthodox nature of their infiltration platforms (i.e., marine-based) and a consequent emphasis on maximising shock effect through a well-synchronised assault.

The Selous Scouts were unusual in that they recruited both from other units of the Rhodesian Army, as well as directly from the civilian population. For a unit that worked in sensitive cross-border operations, civilian recruitment was a risky move, but the danger of security breaches was contained by an elaborate screening process. First, members of the unit would travel to a recruiting area, wearing mufti and masquerading as unemployed drifters. Over friendly chats, they would hint to local men that serving as soldiers might be worthwhile. They would then pretend to apply to the army, bringing along with them a few genuine applicants. All would be interviewed,

[8] Plaster, *SOG*, p. 308.

and those among the first-timers who were thought useful, would be accepted for a six-month basic training course. This was only a tougher form of infantry training, but still had a 33 per cent fail rate. Those who made it through joined volunteers from other army units, and together they all would attend a seventeen-day selection course to enter the Selous Scouts proper. From this point, harassment of the trainees would be unrelenting, and would include forcing them to gather food on their own from the jungle, in their free time. Meat, when it was provided, came from animal carcasses that had been left to rot in jungle heat for at least five days. The dropout rate during the selection course was 85 per cent. Those who made it through would be sent on a fourteen-day course in bushcraft, and among these, the particularly intelligent ones would be sent on a further twelve-day course in surveillance and agent-handing.[9] This last group would be the crème de la crème—the operators who would infiltrate guerrilla strongholds by posing as guerrillas themselves, gather information on rebel movements, and either interdict them directly or call in reserve troops.

Toughness is No Indication of Competence

There is, unfortunately no data on how effective each type of recruitment strategy is in sourcing the best manpower. Israeli SOF units, for instance, draw their men directly from the cohort that is conscripted into the armed forces every year. Applicants to Sayaret Matkal, considered the most prestigious of all Israeli special units, are volunteers from the conscript intake. Many are recruited through informal contact networks, which include family members. The Russian Spetsnaz, on the other hand, arbitrarily draft in men who are thought to be suitable from a particular year's conscript pool. The candidates are observed during the basic training that all Russian conscripts undergo and if thought to have potential, are ordered

[9] Reid-Daly, *Pamwe Chete*, pp. 146-153.

to report for special duties. Refusal is not an option. They are put through a brutal hardening process that leaves only the toughest still on the training programme. Realism is emphasised to a degree that is perhaps not viable in more liberal states: For escape and evasion training, candidates are dressed in convict uniforms and tasked to travel on foot through a city where local police have been told that a bunch of dangerous criminals has escaped from custody.[10]

Considering the variety of methods used by SOF to attract/compel the best officers and men to serve, there is little in common other than an institutional desire to find the most resolute and physically capable soldiers. Making comparisons about which particular unit is best would be a waste of time.

The 'intrinsic strength/acquired skill' model offers many advantages: It keeps training overhead costs low, provides well-conditioned and motivated manpower, and helps an army measure up to the SOF capabilities of its likely adversaries. However, there is a significant downside: The model places so much emphasis on physical toughness that it is self-limiting in the type of volunteers it attracts. In armies where service in SOF is optional, many see such units as the preserve of social rejects (Israel is, of course, a notable exception). Even if the image is wholly inaccurate and undeserved, the fact remains that SOF tend to be lampooned as a refuge for those officers and men who were unable to fit into a regular military unit. The strain of SOF duties, which is inescapable, also leads to them being associated with slightly diminished intellects. Overall, the kind of trainee who is most likely to make it through an intrinsic strength/acquired skill programme will not be a scholarly type. He is, instead, likely to come from a deeply conservative background, and will be rooted in a small-town or village mindset. This mindset

[10]Kenner, Ron (pseudonym), 'Soviet Army Spetsnaz: The Red Elite from the Hysteria of the 1980s to the Present', *Journal for Intelligence, Propaganda and Security Studies*, Vol. 4, No. 1 (2010), p. 97.

is what he will need to put up with the bodily stress and verbal barbs that instructors will throw at him. Yet, it will also restrict his ability to operate in a fast-paced and vaguely-defined urban operational environment, which is the natural realm of counterterrorism. More than strength, what is needed in such contexts is linguistic skill and personal innocuousness—the gift of the gab.

The Acquired Strength/Intrinsic Skill Model

Aptitude testing lies at the core of an alternative training model: The acquired strength/intrinsic skill paradigm. Though it has come to be adopted in certain American and British SOF units, these tend to possess niche expertise, primarily in intelligence gathering. Unlike direct action units, SOF intelligence units have particularised training regimes and distinct area specialisations. This allows them to recruit the best quality manpower out of a smaller pool of candidates, and invest time and effort in individually preparing these trainees for the rigours of special operations.

The origins of acquired strength/inherent skill training date to the German Wehrmacht. As Hitler's Germany was preparing to go to war in 1939, a need was perceived for a surgical strike capability deep behind enemy lines. Small teams of infiltrators, working in disguise, needed to get past the enemy's front line defences and attack critical infrastructure in his hinterland. The resultant shock effect would paralyse his army, even as the Wehrmacht launched a coordinated offensive. The task of raising such teams was assigned to the Abwehr (German Military Intelligence).

Scouring army personnel files, the Abwehr put together a company-sized force of men who could convincingly pass off as natives of the countries that Germany would invade first. Under the cover name Bau-Lehr Bataillon 800, this unit formed the nucleus of what later become famous as the Brandenburg Division. Having the luxury of calling upon personnel from the entire Wehrmacht (a conscript army), it concentrated on identifying soldiers with linguistic skills and experience of living abroad. Only then did it attempt to mould them into effective rear-area fighters.[11]

Throughout the war, the basic method remained the same: Brandenburg troops would wear either mufti or enemy uniforms and bluff their way into hostile territory, relying on sheer audacity coupled with detailed pre-mission intelligence briefings, to get close to their targets. Their favourite disguise was non-commissioned officers' uniform. The NCO was not as junior as an ordinary soldier who would never be left by higher-ranking personnel to just go about his business, and not as senior as a commissioned officer who would be expected to show familiarity with the latest military regulations. NCOs were inconspicuous and yet trusted: A perfect combination for an infiltrator.

Being able to blend in with the enemy was critical to deciding a candidate's suitability to serve in Wehrmacht special operations. His fitness level and prior service record was secondary. All that was initially required was to give him a clear idea of the risks involved: Each trainee was warned that if caught by the enemy, he would be summarily executed as a spy for having fought while in disguise. Each was told that if he wished to withdraw from an operation, he was free to do so, provided he gave adequate notice to his immediate commander. Failure of nerve during an actual mission would be considered cowardice and dealt with harshly, as permitted by military law.[12]

British Counterintelligence Training in Northern Ireland

The Brandenburgers played a crucial role in securing Germany's spectacular initial successes during World War II. Their tactics, though underhanded, were copied by all sides but never with quite

[11] Weale, *Secret Warfare*, p. 49.

[12] Lefevre, *Brandenburg Division*, p. 57.

the same level of effectiveness. Following the war, the British army adopted similar principles in Northern Ireland when it created the Force Research Unit and 14th Intelligence Company. Since the aim of these units was to fight terrorism, as far as the British were concerned the means used were less important than the results obtained, and the results were impressive. Within ten months of being established in late-1972, 14 Int helped the army capture sixteen high-ranking IRA operatives.[13] Not to be left behind, the FRU pulled off similar coups of its own, the most spectacular of which came in 1978, when it recruited the deputy head of the IRA's mole-hunting unit, to work as a mole.

Training for these two units differed from that of the SAS, in that it was more taxing on a candidate's interpersonal skills and street-smartness. Typically, it began with volunteers responding to an announcement for unspecified 'special duties' in the army. Only serving soldiers could apply. Unusually for special operations units, women were encouraged to join 14 Int, since they provided excellent cover during urban surveillance operations: bystanders would naturally assume that a young couple lingering at a public spot were romantically involved, rather than being covert operatives. Moreover, 14 Int and FRU did not place as much emphasis on physical strength as did more mainstream SOF units. Instead, the training focus was on building stamina and resourcefulness.[14]

All volunteers were pre-screened at their parent units before being accepted for training. At 14 Int, there was a three-week selection course, similar to that of the SAS but involving more psychological shocks caused through expectation disruption. This was

[13] Hughes, Geraint, 'The use of undercover military units in counter-terrorist operations: a historical analysis with reference to contemporary anti-terrorism', *Small Wars & Insurgencies*, Vol. 21, No. 4 (2010), pp. 572-573.

[14] McNab, Andy, *Immediate Action* (London: Corgi, 1995), p. 360.

followed by a week's battle inoculation training, designed to give candidates a taste of operating conditions in Northern Ireland. Finally, there were four months of advanced training in unarmed combat, clandestine photography, foot and mobile surveillance and counter-interrogation. Emphasis was given to enhancing situational awareness and to reflexive shooting with concealed weapons. Since a 14 Int operator could not carry much firepower, the volunteer needed to appear defenceless, and thereby tempt an attacker into exposing himself before firing. In 1981, one operator was saved by his training when his car was ambushed by four IRA gunmen—so fast was his reaction that he was able to shoot three of them before escaping.

The FRU did not have a 'selection' component to its training curriculum—candidates either passed or failed, based on their test performance. There was no possibility of second attempts. To gauge their aptitude for agent-handling, each candidate was put through three days of preliminary testing, wherein they had to elicit personal information from unsuspecting members of the public. Details gathered were cross-checked against police and other government records. 80 per cent of all candidates would fail at this point, thus ensuring that the remainder (usually about ten) who showed a natural talent for winning other people's trust, were the ones accepted to the actual training. Of these, more than half would either fail during the eight-week course, or voluntarily drop out. Thus, at the end of the training process, less than 10 per cent of those initially tested would make it through to join the unit.[15]

On the surface, there are some similarities between the acquired strength/intrinsic skill process used by the Brandenburgers, FRU and 14 Int and the intrinsic strength/acquired skill processes used by the SAS, Delta Force and SEALs. Failure rates are about the same (80-90 per cent) and physical testing is used to weed out the unmotivated

---

[15] Ingram and Harkin, *Stakeknife*, p. 29.

from the genuinely committed.[16] However, there are important differences: Training time is much lesser in the acquired strength/intrinsic skill model. It takes about six months to fully condition an operative to go out into the field, as opposed to a minimum of two years for soldiers who go through SAS-style training. In part, this is because units like 14 Int and FRU, or even the Brandenburgers before them, are proficient only in certain locales. While SAS troopers have to be ready to operate anywhere in the world, 14 Int and FRU can operate only in Northern Ireland, because their tactical procedures are optimised for this context. Similarly, the Brandenburgers who were deployed against Poland and on the Russian Front during World War II, would have provided no added value to Wehrmacht offensives, if used to fight against France or Belgium.

Combining the Two Training Models

Perhaps the only SOF unit which successfully combines the strengths of both training models is the US Army Intelligence Support Activity (ISA). The unit approaches prospective members on the basis of their individual skills, and only if they are found temperamentally suitable, admits them to its training programme. Initial contact is made on the basis of personnel searches conducted on the US Army's computerised roster. Anyone with an intelligence or ethnic minority background is deemed of interest, particularly if they come from regions where the US has long-standing interests, such as Latin America, the Middle East or Southeast Asia. Dual passport-holders are considered major assets, since they can travel to hostile countries without requiring forged documents.

[16] In contrast, the failure rate for Indian Special Forces is less than 50 per cent. Training is conducted on the British intrinsic strength/acquired skill model, thus making for a tough but unsophisticated soldier incapable of operating in denied territories without massive conventional support. Shekhar, Chandra, 'Special Forces in India: Doctrinal Issues and Employment Concepts', in Oberoi, Vijay ed., *Special Forces: Doctrine, Structures and Employment Across the Spectrum of Conflict in the Indian Context* (New Delhi: Knowledge World, 2006), p. 155.

Testing for the ISA is known as 'Assessment and Selection' or A&S. A typical test would involve leaving a trainee in the middle of a desert somewhere in the western US, with no provisions or communication equipment. The trainee would not only have to stay alive against the elements, but gather answers to a pre-assigned list of intelligence questions, using entirely his own resources. Upon completion of this task, he would be deprived of sleep and dropped off in an unfamiliar city, where he would have to follow a designated person without getting noticed himself by counter-surveillance teams. The candidate who successfully completed these tests would then be put through commando-type training, such as parachuting, weapons handling, escape and evasion and counter-interrogation. As with the Selous Scouts, there would also be a course on agent-handling.[17]

Good Training is Time-And Resource-Intensive

Naturally, investing so much effort in a single trainee is costly, and raises the question of whether the returns are worth the resources consumed. However, the same can be said for all special operations training, including in prestigious units like the SAS. The added value of SOF comes from being able to operate in small, tactically autonomous teams that require minimal extraneous support.[18] Such self-sufficiency does not come automatically—it is cultivated by individual preparation. On average, an eight-man reconnaissance team carries around four times the volume of firepower as a comparably-sized infantry section.[19] Having the strength to haul combat gear over rough terrain requires months of physical toughening. Under the dominant paradigm—intrinsic strength/acquired skill—

[17] Smith, Michael, *Killer Elite: The Inside Story of America's Most Secret Special Operations Team* (Weidenfeld & Nicolson), p. 40.

[18] Beckwith, Charlie A, *Delta Force: The Army's Counterterrorist Elite Unit* (New York: Avon, 1983), p. 112.

[19] Plaster, *SOG*, p. 153.

such toughening is necessary even before an individual applies to join an SOF unit. The time spent preparing for commando training detracts from his regular duties with the parent unit.

Once a candidate passes the selection phase—in either training model—and enters advanced training, overhead costs will be high in any case. Ammunition expenditure is an obvious example: On average, Delta Force trainees fire 500 rounds daily, while the SAS, not being as well-resourced, makes do with 5000 rounds per day for all trainees combined. The US Navy SEALs are known to use 'simunitions'—9mm paint-loaded rounds that can be fired by pistols and sub-machine guns of the same calibre, allowing for opposed close-quarter battle (CQB) shooting under realistic conditions. Readers can compare this with the plight of police commandos in some Indian states, who can count themselves lucky if they receive more than 150 rounds *per year* for training purposes.[20] Little wonder that the police response to the 26/11 attacks was so poor—even allowing for the unprecedented sophistication of the attack, first responders were simply outgunned by much better trained adversaries.

This leads to an uncomfortable reality: Terrorists focus on combat training. Initially, Al Qaeda projected itself in the international *jihadist* fraternity as a provider of high-quality tactical instruction.[21] The bulk of its trainers were ex-members of the Egyptian security forces, which gave the group much-needed credibility. Lashkar-e-Taiba (LeT) copied this marketing approach, bringing in former members of the Pakistani Special Services Group to work as trainers.[22] The very concept of *fidayeen* attacks—assaults which would unfold so rapidly that police first responders would not be

[20] Sharma, LN and Kumar, Umesh, 'A Need for a Cogent Counter-Terror Doctrine', *Eternal India* (2009), p. 90.

[21] Bergen and Cruickshank, 'Revisiting the Early Al Qaeda', p. 12.

[22] Zaidi, S Hussain with Bhatt, Rahul, *Headley and I* (New Delhi: HarperCollins, 2012), p. 80.

able to intervene—was thought up by a former NCO from Zarrar Company, the SSG's counterterrorist team.

Matching Training to Tasking

Against such adversaries, the Indian state has sent brave but ill-equipped troops who are labelled 'commandos' but possess little of the specialised expertise required of them. Whatever successes units such as the National Security Guards and Army PARA (Special Forces) battalions have produced, have been the result of individual courage and ingenuity, of which Indian soldiers are never lacking. Systemic quality is not assured however, owing to lack of a standardised training programme and institutional support to special operations forces. Service in an SOF unit is still seen in many quarters as a getaway option for misfits unhappy in their parent units. Testing is based on tradition rather than practical sense. For example, the National Security Guard (NSG) regularly conducts five kilometre runs in order to maintain troop fitness levels. Yet, as army officers attached to the force have commented, this is a less useful way of upholding fitness standards than advanced training in martial arts/unarmed combat. Personnel working in counterterrorism need to be capable of swiftly and aggressively neutralising a threat in their immediate vicinity and would not have to run five kilometres just to close in with it.

The mismatch between training and tasking is a result of both institutional pressures as well as a desire to economise. For a long time, Israel assigned the task of hostage rescue to Sayaret Matkal, which as a military reconnaissance unit, was unsuited to the role. The unit had a mixed record of successful hostage interventions, which by most standards, is tantamount to having a bad record. Its personnel only had four weeks of specialised counterterrorism training. Being conscripts serving tours of fixed length, they tended to leave just as they were reaching peak efficiency. After recognising that hostage rescue missions required a full-time unit to undertake them, Israel created the Yamam unit within its border guards. Operators in this unit received twelve months of training, of which four were dedicated exclusively to resolving hostage situations. They also signed on for much longer tours of service, thus helping the unit build an institutional memory through their collective experience. However, so powerful was the Sayaret Matkal lobby within the Israeli command structure, that Yamam rarely got a chance to deploy—the army unit continued to 'hijack' hostage rescue missions for itself.[23]

A comparably lopsided policy exists in the NSG. The cutting edge of the force is concentrated in just two units, the Fifty-first and Fifty-second Special Action Groups, which are 100 per cent Army. The less well-trained Special Ranger Groups draw their manpower from central paramilitary forces and are only meant to provide cordon security at a terrorist-hit location. Yet, army personnel seconded to the NSG serve tours averaging three years, while paramilitary personnel serve up to five years.[24] This is unproductive, since the elite units that actually would play the lead intervention role during a crisis have a higher turnover rate, and therefore, must waste time training new personnel. Furthermore, it is disruptive upon officers' and troops' family life. For the sake of professionalism and building a strong sense of unit cohesion, it would make sense to allow army personnel in the NSG to serve longer tours.

The Indian Context

Indian special operations forces deployed against Pakistan need to be able to pass off as locals. Even allowed for the common Punjabi and Hindustani (Urdu/Hindi) languages, this is an extraordinarily

[23] Pedahzur, Ami, *The Israeli Secret Services and the Struggle against Terrorism* (New York: Columbia University Press, 2010), p. 51-52.

[24] Anonymous comment, cited in Yadav, VS ed., *Employment of Special Forces: Challenges and Opportunities for the Future* (New Delhi: Vij Publications, 2012), p. 112.

difficult feat to achieve. In densely populated locales, a person who acts, ever so slightly, as though he is an outsider is immediately identified as one by the local inhabitants. Personnel who are sent to Pakistan for operational purposes, therefore, need to be mentally conditioned for surveillance and countersurveillance tasks, rather than for reconnaissance. With the availability of tactical Unmanned Aerial Vehicles (UAVs) and satellite imagery at the forward command level, SOF are increasingly becoming outdated as instruments for gathering topographical intelligence. Instead of strength, what they need is the ability to blend in with the enemy and independently manoeuvre into strike positions without being observed.

The transformative effect of new technology cannot be underestimated: Much of the conditioning that SOF troops are put through is to enable them to work as 'eyes on the ground'. Since World War II, the Anglo-American model of SOF training has emphasised operating in rural environments, because that is where large military forces are likely to engage in head-on clashes. However, the dominant trend in warfare has been somewhat different: Instead of conventional symmetry, what characterises modern battles is their pointed asymmetry. Rarely, if ever, does the manifestly weaker side engage the stronger on terms that favour the stronger side. In general, and especially in the fight between terrorists and governments, this means that one side will try to fight in urban terrain and hide amongst the civilian population, provided the other side shows a preference for manoeuvre warfare in open spaces and tries to avoid inflicting collateral damage. The Indian Army is the second type of fighting force.

Besides total familiarity with the public administration machinery in Pakistan, SOF units need to be trained in evasive driving to the standard normally reserved for VVIP protection. A surveillance/ assassination team that needs to get out of Pakistan cannot count on aerial extraction—it will have to use its own resources and terrain knowledge to navigate back to Indian territory. This can only be made possible if each special operator is compelled to familiarise himself with the border road network in the Pakistani sector immediately facing his launching pad. Training must encompass such mapping of the physical and infrastructural terrain. Instead of having a general purpose SOF unit to carry out direct action missions in enemy territory, India should create intelligence-oriented units which can also serve in a direct action capacity when necessary. It should, therefore, adopt the acquired strength/intrinsic skill model of SOF training, and channel all such personnel towards special operations, who possess the necessary linguistic and mental aptitude.

Part II

Threat and Response

This is the era of the proxy agent and surrogate combatant, an era of strange alliances, of double and triple agents, of small and massive deceptions, of subagents and subcontract agents in which even the proxies have proxies.[1]

—Paul Balor (US mercenary soldier)

[1] Balor, Paul, *Manual of the Mercenary Soldier: A Guide to Mercenary War, Money, and Adventure* (Boulder: CO: Paladin Press, 1988), p. 61.

5

Pakistan's Proxy War

This chapter explains how the Pakistani Inter Services Intelligence (ISI) Directorate has conceptualised its urban paramilitary operations in India. It suggests that Indian efforts to mobilise the West against Pakistan are doomed to failure, owing to differences in how India and the West perceive state-sponsored terrorism. The chapter also delineates the role of *jihadist* groups such as the Lashkar-e-Taiba (LeT) and Indian Mujahideen in this undeclared war.

Pakistani support to terrorism is driven as much by ideological motives as strategic calculation. The two influences coexist within an operational design that satisfies the ISI's sanguinary instincts while exploiting Indian vulnerabilities to Pakistani advantage. First, the chapter shall outline three schools of thought that motivate Pakistani covert action in India. It shall demonstrate how such covert action can extend to outright sponsorship of terrorism, owing to the inability of Western analysts to appreciate contextual particularities that only apply to India. Thereafter, the chapter shall describe how the ISI came to support urban terrorism since the 1990s, and why India failed to combat the threat effectively. Finally, it will provide a brief outline of trends in cross-border terrorism.

Rationales for Subversion: Irredentist, Salafist and Instrumentalist

Three distinct strands of reasoning motivate Pakistani covert interference in Indian domestic affairs. First, there is the Irredentist school, which claims that Pakistan as a homeland for Indian Muslims has automatic rights over Jammu and Kashmir (J&K). India, according to this interpretation, is not a viable nation state and hence, its claims to sovereignty, whether in J&K or elsewhere, are false. Second, there is the Salafist school, which draws upon Arabian *jihadist* narratives to emphasise the unity of the Islamic world against unbelieving adversaries. Lastly, there is the Instrumentalist school, which focuses on the practical benefits of attacking India covertly, irrespective of Irredentist or Salafist logic.

IRREDENTIST MOTIVES

The Irredentist school is an Indo-Asian version of the exclusivist logic which tore Europe apart during the Thirty Years War (1618-48) and led to the formation of nation-states defined by Protestant and Catholic demographics. According to such logic, the only real nation in the Indic landmass is Pakistan, a state purified of centrifugal forces by the cultural bond provided by Islam. The rest of the Indian subcontinent is only a geographic expression, like 'Europe' or 'Africa'. It is not a country but a hotchpotch of many nations that are yet to attain their sovereignty.

In the years after India's Partition, such views were particularly common among officers of the Pakistan Intelligence Bureau. One officer, named Qurban Ali, propounded a theory that postcolonial India in its present unified form was a historical aberration whose disintegration was inevitable. His assessment was adopted across the entire national security architecture of Pakistan. Thus, from the early 1950s, the Pakistan Intelligence Bureau (IB) began infiltrating covert operatives into J&K, with the aim of instigating a rebellion.[1] Its subversive activities were motivated not just by the religion-based claim that Pakistan had advanced on the Muslim-majority region, but also by a belief that India in general was not a legitimate state. Rather, this state was an artificially constructed colonial inheritance, held together against the will of its multi-ethnic, multi-lingual and multi-religious peoples by a repressive federal government. Its breakup would be an act of deliverance for them.

Thus, the Pakistan IB, and not the ISI, was the original author of cross-border terrorism in India. Unlike the ISI, which was founded in 1948 to coordinate military intelligence, the Pakistan IB already had extensive knowledge of covert operations. Its British-run colonial ancestor had been partly responsible for fuelling Hindu-Muslim tensions, and thus, creating Pakistan in the first place. All that the Pakistan IB had to do was to build on this expertise and use it to undermine the Indian state's claims to sovereignty after 1947. The agency remained crucial to covert action against India until the 1970s, when the Soviet-Afghan War and domestic turmoil in Pakistan boosted the ISI's profile. Thereafter, the latter agency took over responsibility for covert operations.

SALAFIST MOTIVES

The Salafist school of subversion is an Arabised version of the Irredentist school, since it, too, denigrates the Indian state as illegitimate. It differs, however, in its emphasis on the historical right of Muslims to rule India. Unlike the Irredentist school, which suggests that India is an inherently fractured entity, the Salafist argument is that India is a breakaway region within Greater Pakistan.

---

[1] Swami, Praveen, *India, Pakistan and the Secret Jihad: The Covert War in Kashmir 1947-2004* (New York: Routledge, 2007), p. 46.

It asserts that a conspiracy between the West and Hindus has deprived Muslims of their ancestral right to govern the Indian sub-continent.[2] The long-term aim of Salafist ideologues is to re-establish Islamic rule over the entire Indic landmass. It has been popularised by decades of institutional contact with similarly radical Pan-Islamist militants in the Persian Gulf states, particularly Saudi Arabia.

Part of this Arabising trend was facilitated by the Pakistan Army. During the 1970s, the army reinterpreted the Quran to justify offensive operations against India in peacetime. In 1979, it authorised a study called *The Quranic Concept of War*, arguing that terrorism, by paralysing an enemy society, could also paralyse its military and make lengthy force-on-force campaigns unnecessary.[3] The study stated that the Clausewitzian model of warfare was not applicable to Pakistan. Instead of war being governed by policy, policy must adapt to the dictates of war. This was because any decision to wage war was made by God, and policy must not be allowed to interfere with that decision.

The implications of such logic need to be considered. If applied to covert operations, they suggest that rather than have a clear-cut endgame, the ISI must first cause maximum disruption within enemy society. Only then should a political objective be decided upon. This would explain why the ISI tends to shoot first and strategise later. As one British writer has noted, despite its emphasis on covert operations, the agency has done a bad job of protecting or serving Pakistani national interests.[4]

The Arabisation of Pakistan occurred in the aftermath of the 1971 War, in which the country suffered a catastrophic defeat to India.

---

[2] Nayyar, AH and Salim, Ahmed, *The Subtle Subversion: The State of Curricula and Textbooks in Pakistan* (Islamabad: Sustainable Development Policy Institute, 2006), p. 81.

[3] Myers, Joseph C, 'Review Essay: The Quranic Concept of War', *Parameters* (Winter 2006-07), pp. 116-117.

[4] Johnson, Robert, 'Pakistan's Inter Services Intelligence', in Davies, Philip HJ and Gustafson, Kristian C eds., *Intelligence Elsewhere: Spies and Espionage Outside the Anglosphere* (Washington, DC: Georgetown University Press, 2013), p. 135.

In the absence of military might, it built up its covert capabilities. After 1971, the ISI set up hundreds of 'sleeper cells' across India.[5] Their job was to remain deep under cover until another war broke out, whereupon they were to engage in sabotage. It was these cells which were subsequently used to import Salafist ideas into India from the early 1990s onwards.

INSTRUMENTALIST MOTIVES

Lastly, there are instrumentalist reasons for engaging in covert action against India, even if Irredentist or Salafist motives are disregarded. Cross-border terrorism imposes tremendous financial and societal costs on the Indian state. Since this state is viewed as a sworn enemy of Pakistan, a zero-sum situation exists wherein any loss to India is a gain for Pakistan.

Covert interventions in India's domestic conflicts provide two immediate advantages:

- They exacerbate the country's internal security situation and tie down large numbers of troops in policing duties. Such troops, as per the Pakistan Army's view, would normally be arrayed along the border in a threatening posture. Therefore, by supporting rebel groups, the ISI averts a dangerous military imbalance.[6]
- Such interventions increase scope for intelligence collection by human assets, since they weaken the emotional bond between the Indian state and its citizens. Official overreaction within India to armed rebellion often alienates entire communities, thus providing the ISI with a fresh manpower pool to recruit agents from.

---

[5] John, Wilson, 'Target practice in the country', *ORF Analysis*, 18 January 2006, http://orfonline.org/cms/sites/orfonline/modules/analysis/AnalysisDetail.html?cmaid=2701&mmacmaid=1042, 30 October 2013.

[6] Raman, B, 'Pakistani sponsorship of terrorism', *South Asia Analysis Group*, Paper No. 106, http://www.southasiaanalysis.org/paper106, 30 October 2013.

In addition, there are two long-term advantages, which make covert action attractive:

- In the long run, covert action might lead New Delhi to conclude that it cannot achieve domestic prosperity unless it concedes Pakistani demands on J&K. The financial damage caused by violent subversion, in the form of property losses, extra security expenditure and wasted business opportunities, could force the Indian government to choose between development in the rest of India, or possession of J&K. Either way, Pakistan would be insulated from any negative fallout. Its nuclear arsenal would make military retaliation impossible, while its usage of local proxies would mitigate international criticism. Rather than blame the perpetrator, foreign governments would most likely castigate the victim for failing to provide good governance to its citizens.
- Also, Indian officials would have a hard time proving Pakistani involvement with rebels, since the subversive infrastructure required for both insurrection and espionage is identical. This is a unique strength that Pakistan enjoys in its covert war with India—it can always claim that New Delhi confuses spying (an illegal but accepted, component of statecraft) with sponsorship of terrorism (which is universally condemned). India is not a military ally of the West, but is an object of public hostility in Pakistan due to both Irredentist and Salafist narratives. Covert action against it is, therefore, a low-risk/high-return endeavour.

Analysing State-Sponsored Terrorism: Comparing Western and Indian Models

Normally, Western analysts conceive of state-sponsored terrorism as a specific incident or series of incidents, where evidence can be gathered directly implicating a head of state in a terrorist act.

Thus, it is not enough to prove which country a terrorist attack cell comes from. Investigators also have to prove that its members received support from official agencies of that country and that such support was authorised by the highest decision-making authority.[7] As the West perceives it, a state-sponsored terrorist incident shows a discrete command chain:

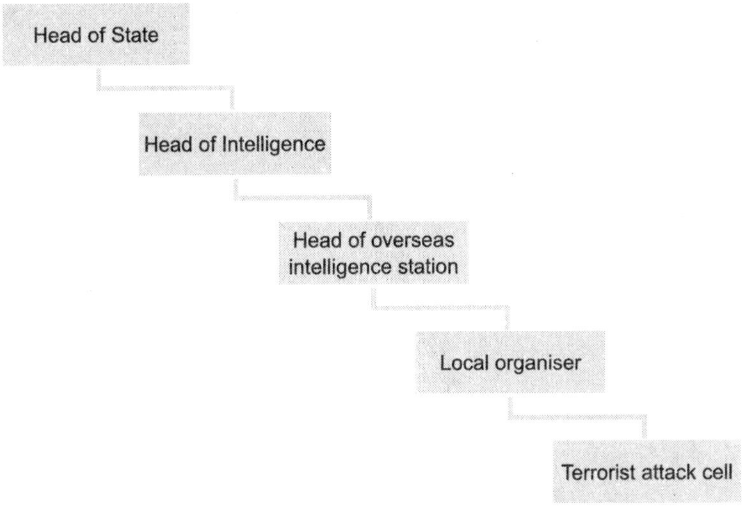

Graphic 1: How the West perceives state-sponsored terrorism

Since state sponsors of terrorism do not share borders with Western countries, an attack upon the West requires considerable planning and preparation. In the process, an electronic trail is created. After the event, investigators can follow this trail back to the overall controlling figure. Typically, it would start with a rogue head of state instructing his intelligence chief to carry out an attack on Western interests. The intelligence chief would in turn be linked to an embassy-based senior intelligence operative in the West, who would arrange logistics for the attack through a local intermediary.

[7] Conversation with top US counterterrorism expert, September 2010.

Finally, this local intermediary would be in direct contact with the actual terrorist cell that launches the attack.

What India faces from Pakistan is a different type of threat, consisting of a constant stream of small-scale hostile actions carried out on a near-autonomous basis, rather than a few big ones ordered by top decision-makers. It is a fusion of espionage and terrorism, where the two functions are kept apart at the operational level but are centrally coordinated by one body.

The Pakistani covert action command structure is as given below:

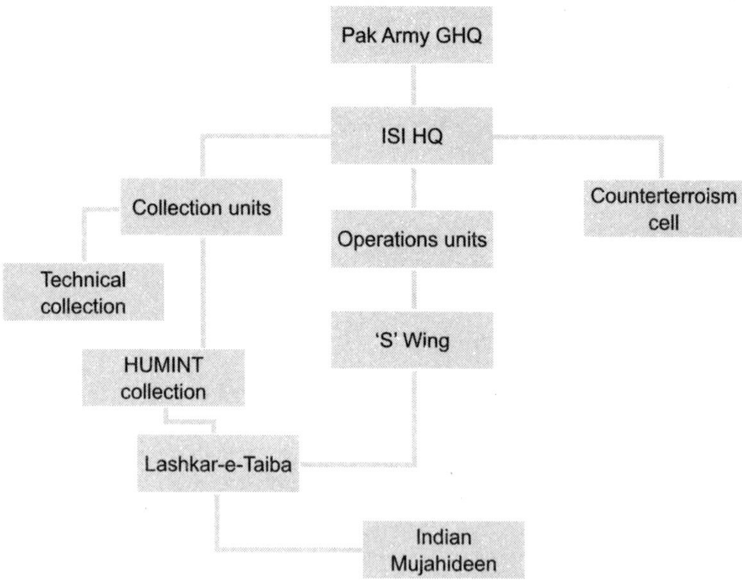

Graphic 2: How the Pakistan Army and ISI sponsor terrorism

The graphic shows how compartmentalised the ISI's relations with terrorist groups are. Directives routinely passed down from Army General Headquarters are refined into intelligence targeting priorities at ISI Headquarters. While technical and human intelligence units focus only on collecting and analysing information, operations units engage in covert action to further Pakistani interests. One such unit, structured such that it works separately from the rest of the agency, is the counterterrorism cell. Its job is to cooperate with US intelligence agencies in tracking down Al Qaeda suspects. Another operational unit, better integrated into the mainstream work culture of ISI, is 'S' Wing. Consisting of retired and re-employed soldiers, its job is to support *jihadist* groups that have convergent interests with the Pakistani state. Not surprisingly, the counterterrorism cell and 'S' Wing often work at cross-purposes. Given the much greater institutional clout of the latter, it remains the dominant faction. 'S' Wing is responsible for handling the LeT, which in turn serves as a mentor and intermediary for India-based *jihadists* (the Indian Mujahideen).

Taking the example of the 2008 Mumbai attack, it appears that the command chain was as follows: In late 2007/early 2008, a decision was made at very senior levels within ISI Headquarters to facilitate a major terrorist attack on India. Its implementation was turned over to 'S' Wing, which was already conducting tactical reconnaissance through human assets in India. Some of these assets were Indian Mujahideen members used exclusively to support *jihadist* activities, while others like the Pakistani-American mercenary, David Headley, were tasked to simultaneously collect intelligence on Indian military installations. One of Headley's co-conspirators, Tawahhur Rana, later told US investigators that he was misled into thinking that the Mumbai operation was a regular espionage mission by the ISI.[8] He claimed that since he had acted in good faith as a patriotic Pakistani, he could not be blamed for the terrorist attack. The actual attack was carried out by LeT, with advisory and logistics support

---

[8]'Mumbai terror attack: LeT's Tahawwur Rana confirms Pakistan government, ISI's 26/11 role', *Economic Times*, 13 April 2011, http://articles.economictimes.indiatimes.com/2011-04-13/news/29413739_1_rana-and-headley-tahawwur-hussain-rana-isi, 8 November 2013.

provided by at least six serving ISI officers (five of them army majors).[9]

Since the 9/11 attacks, the ISI has helped disrupt terrorist threats to Western homelands while working against Western interests in South Asia. In the process, it has not relented on its core policy of targeting India, and has not been pushed to do so by Western governments. Owing to the agency's seemingly contradictory stances, Western analysts have difficulties in identifying where its loyalties lie. They have suggested that a powerful 'rogue' faction exists within the ISI and continues to maintain contact with *jihadist* groups. From an Indian perspective, however, the problem is clear: There can be no 'rogue' faction within the ISI because there is no 'friendly' faction to begin with. Even if some departments are engaged in the routine task of collecting information for political or military purposes, others are waging a covert war on India through violent proxies. Both elements work within a common organisational structure. Thus, as far as Indian analysts are concerned, the ISI and not merely LeT or Indian Mujahideen, is a large part of India's terrorist problem.

The three motives for ISI support to terrorism—Irredentist, Salafist and Instrumentalist—are like concentric circles drawn around a bull's eye. They all have a common focal point: India. Whether cross-border terrorism is seen as a nation-building project, or a religious struggle to reclaim a stolen realm, or an opportunistic thrust against a vulnerable enemy, it has always been an integral part of the ISI's job description as far as New Delhi is concerned. The next section shall narrate how support for highly provocative, large-scale attacks in Indian cities became part of this description in the 1990s.

Anatomy of a Conspiracy

The year 1991 marked a turning point in the ISI's operational culture. Until then, the agency had mainly concentrated on supporting

[9] Conversation with a US Central Intelligence Agency official, June 2011.

rural insurgencies. In India, it had provided financial, logistical and intelligence support to the Naga, Mizo, Sikh and Kashmiri separatists, and in Afghanistan, it had organised a guerrilla war against the Soviet Army. The collateral damage caused by such activities was felt mainly in villages and remote peasant communities. Their political impact was minimal. Although this helped Pakistan keep bilateral tensions with India and Afghanistan at a manageable level, it also limited the effectiveness of covert action.

By 1991, the ISI had started to perceive the shortcomings of a paramilitary covert action strategy that focused only on rural targets. It had been supporting Sikh separatists for over ten years, to little effect other than several thousand dead Indian peasants. New Delhi had responded to separatist violence by saturation deployments of police and paramilitary personnel, thereby containing the damage caused. From this pattern, the ISI deduced that a village-based rebellion would do no irreparable harm to the superstructure of the Indian state. It needed to create a dispersal effect.

An ISI major named Waqar Ahmed came up with an innovative suggestion: Why not coordinate the activities of Sikh and Kashmiri insurgents and help them build an attack capability in Indian cities? Starting in September 1991, he organised strategy sessions between leaders of both separatist movements, who were already based in Pakistan.[10] He also enlisted the support of the Jamaat-e-Islami (JeI), Pakistan's largest Islamist political party, which was suffused with ideological animosity towards India. Working with the propaganda department of JeI, the ISI recruited covert operatives to function as linkmen between Sikh and Kashmiri terrorist cells in Indian cities. Its existing operational assets in northern India were also tasked to assist coordination between the two groups of separatists.

---

[10] *Review of Terrorist Activities in Punjab for the month of September 1991,* compiled by the Office of the Additional Director-General of Police Intelligence, p. 42.

Ahmed's plan for a paramilitary offensive against Indian cities was initially code-named K2, since it symbolised cooperation between Khalistani (Sikh) and Kashmiri insurgents. Once Hindu-Muslim tensions across India worsened in 1992, over the status of the Babri Masjid, another dimension was added: Islamic militancy in areas other than J&K. The ISI recruited radical activists of the Students' Islamic Movement of India (SIMI) to reconnoitre strategic targets such as military installations and large-scale economic facilities. It also compelled the Indian-origin crimelord Dawood Ibrahim to provide logistical support to terrorist operations (by seizing his smuggling fleet in early 1993).[11]

The Mumbai blasts of March 1993 were the first manifestation of the K2 scheme, now modified to K2M (Khalistan, Kashmir and Muslim Militancy). 257 people were killed in eleven synchronised explosions across the city. The blasts aimed to cause maximum loss of life, since they were directed at 'soft' or unprotected targets. According to confessional statements made by captured members of the bomb network (all career criminals working for Dawood Ibrahim), the choice of targets was decided by controllers in Pakistan—a likely reference to the ISI.[12] Arms and explosives recovered from them had Pakistani markings and the detonators used were traced to Pakistani army ordnance factories. Yet, the US, when presented with this evidence, insisted that just because the bombers had trained in Pakistan and had sourced their hardware from its territory, the Pakistani state could not be blamed. According to one Indian source, the US government even obstructed investigations into

---

[11] Lele, Ajey, 'A Shift from Crime to Terrorism: Assessing D-Company', *Society for the Study of Peace and Conflict (SSPC) Paper 4* (December 2005), p. 7, http://www.sspconline.org/sites/default/files/RP_04.pdf, 31 October 2013.

[12] Zaidi, S Hussain, *Black Friday: The True Story of the Bombay Bomb Blasts* (New Delhi: Penguin, 2002), p. 73.

Pakistani involvement in the blasts conspiracy, in order to protect its traditional ally.[13]

At the time, what happened in Mumbai seemed to be a one-off event. The blasts had been preceded by horrendous rioting between Hindus and Muslims, in which the latter had suffered disproportionately heavy losses. Mumbai 1993 could have been dismissed as an isolated case of revenge by an oppressed minority. Yet, intelligence reports at the time warned that the ISI was looking to capitalise on precisely such tensions and build up paramilitary cells within India.

Almost simultaneously, documents captured from Kashmiri separatists indicated that a shift in their targeting policy might be imminent. Attacks on the common public had been assessed as more likely to weaken the Indian state than raids on government forces. Such views had previously also appeared among Sikh separatist groups, before they were crushed by decisive police action. Intelligence reports specifically named the ISI as the originator and driver of the change in terrorist targeting. One 1997 assessment broadly outlined six components in the ISI's strategy of fostering urban terrorism:

- Destabilising the economy through attacks upon transportation infrastructure
- Disrupting commerce through raids on telephone and stock exchanges
- Triggering civil unrest through sabotage of electricity supply grids
- Assassinating political leaders through suicide bombers
- Instigating inter-religious violence through agent provocateurs, and

[13] Rajghatta, Chidanand, 'B Raman, India's seasoned spymaster and trenchant US critic, dies at 77', *Times of India*, 17 June 2013, http://articles.timesofindia.indiatimes.com/2013-06-17/us/40027055_1_bahukutumbi-raman-1993-mumbai-blasts-pakistan, 30 October 2013.

- Ensuring that terrorist groups in India got legal protection through front organisations.[14]

The heavy emphasis on economic targeting is important, since the year 1991, when the ISI developed its blueprint for urban operations, was also the year of Indian economic reform. Whether this played any role in ISI planning at the time is unclear, but in the intervening years, the agency has had a consistent desire to hit economic targets. Its aim is, apparently, to cause structural damage to the Indian state, without regard for civilian casualties, which are merely viewed as collateral damage. From available data, until recently it had not been very successful in steering terrorist groups in this direction, as the latter wanted to focus on targets of political and religious significance.

However, three factors might now be aligning the targeting priorities of terrorists operating in India with those of the ISI. First, there has been a shift in operational perspectives from the local to the global. Import of Salafist ideas from the Persian Gulf states to Pakistan, and thereafter from Pakistan to India via ISI 'sleeper cells' and LeT, has created a sense of Pan-Islamist unity among *jihadists*. Local grievances, which previously motivated radical elements within India, have yielded to a broader narrative that emphasises civilisational conflict. India is portrayed in such discourse as an enemy of Islam, on par with the US and Israel. Any attack on its interests, whether publicly or privately owned, is legitimate. This sets the stage for a terrorist offensive focused on business ventures, especially those involving Western nationals.

Second, the ISI now has an obedient proxy to handle contacts with Indian *jihadists*: LeT. A new layer of deniability has, therefore, come into ISI covert operations. Even if an attack on India is traced to Pakistani territory, it can always be attributed to non-state actors.

---

[14] *Review of Terrorist Activities in Punjab for the year 1997*, compiled by the Office of the Additional Director-General of Police Intelligence, pp. 18-22.

Operatives of 'S' Wing—who function as the ISI's striking arm, have full rein to plan high-impact operations against India. With the LeT leadership, as per credible accounts, being completely submissive to ISI control, the spy agency anyway remains the dominant actor.

Third, there has been a general drift towards attacks on soft targets. Although casualty maximisation can always be accomplished by hitting crowded religious sites, commercial entities such as public transport or marketplaces are good alternatives. Such locations serve the dual purpose of yielding a high death toll, thus ensuring media coverage, as well as hurting the Indian economy, if only symbolically. As terrorist groups use improvised explosive devices (IEDs) far more often than they use *fidayeen* tactics, attacks on soft targets are low-risk. Unless the perpetrators of an urban bombing were to claim responsibility, investigators would be hard-pressed in identifying them.

In a sample of sixty-four terrorist attacks on Indian cities between 1993 and 2011, only seven involved the use of firearms. The remainder consisted of IED blasts, in some cases involving multiple explosions at many sites. A total of 1,610 Indian citizens were killed in these attacks (allowing for a 1 per cent margin of error due to discrepancies in press reporting). In twenty-one blasts, all of which occurred between 2000 and 2011, investigators were not able to conclusively identify the perpetrators as belonging to any particular *jihadist* group.[15] Meanwhile, Pakistan adjusted to the post-9/11 international environment by courting sympathy as a 'victim of terrorism'. It, thus, placed itself in the same moral bracket as India, negating New Delhi's diplomatic effort to rally international opinion against the ISI. To a large extent, India can only blame itself for this situation. Its counterterrorist response, at both policy and operational levels, has been abysmal.

---

[15] Data gathered from the *South Asia Terrorism Portal*, http://www.satp.org/satporgtp/countries/india/database/OR_9-11_majorterroristattacks.htm, 30 October 2013.

Reasons for Counterterrorist Failures

Firstly, the country's leadership has not contested the Pakistani argument (which is quietly being backed by Britain for ulterior motives) that terrorist violence is in some way linked to the J&K issue. It does not aggressively challenge Anglo-Pakistani assertions that dialogue on Kashmir would offer a solution to terrorism, although the facts clearly do not support such an assessment. Between July 2005 and November 2008, there were eighteen terrorist attacks upon India, killing 820 people. During this period, India and Pakistan held a sustained dialogue to resolve their differences. New Delhi refused to deviate from its commitment to peace talks, until the 26/11 attacks forced a temporary suspension. By not insisting that dialogue with Pakistan can only occur in a context free from terrorist threats, the Indian government has allowed foreign pressure to railroad it onto a counterproductive policy.

Secondly, the Indian security establishment has failed to structurally adapt to the threat of home-grown terrorism with external linkages. This adaptation failure is not as much due to negligence of specific intelligence agencies, as it is to the capacity shortfalls of the system within which they operate. The bedrock of the Indian counterterrorist apparatus, the civilian police, is inadequately equipped for the task. Not only is it fragmented across nearly thirty provinces, thus complicating joint operations against terrorist networks, it also lacks manpower to generate 'actionable' intelligence. National intelligence agencies meanwhile, rely primarily on technical collection to track terrorist activity. They do not have the resources to create local informer networks (except in high-priority regions) and struggle even to extract coherent data from local police forces through liaison arrangements.

The shortfall in human intelligence is key to understanding why Indian security forces were not able to prevent bomb attacks by the

Indian Mujahideen during 2005-08. Trained by the ISI in operational security, the bombing networks maintained good communications discipline, making electronic tracking difficult. Each operational component within a network was compartmentalised. Logistics was handled by one set of operatives, IED assembly by another, and actual placing of the bombs by a third set. New recruits were screened for links to the security forces and anyone with friends or relatives in the police was rejected.[16] The terrorists were able to raise a fairly elaborate counterintelligence system to warn of police raids on their safe houses. Since many had previously been active in organised crime gangs, they had contacts with police informers. Some of these informers were won over and used to monitor the police. Not having human assets who could directly infiltrate the bombing networks, security forces could not gain informational dominance for preventive action.

Thirdly, India lacks a counter-subversion strategy, with deadlines and in-built benchmarks to measure progress. It has relied overwhelmingly on democratic institutions to assuage the grievances of marginalised communities. Such institutions may have prevented a full-blown civil war, but the poor quality of their governance has disillusioned many. Seeking 'quick rich' job prospects abroad, many young men have travelled overseas. There they have come into contact with practiced manipulators from the ISI and LeT whose task is to identify and win over emotionally vulnerable Indian migrants. New Delhi has not invested in building a future for these young men, so it has no automatic claim to their loyalty. Furthermore, it has not reactivated its covert action capability, which was shut down on executive orders in 1997. Using this capability, intelligence agencies could have inserted long-term penetration agents into foreign-based terrorist groups and provoked factional disputes. Instead of

[16]Gupta, Shishir, *Indian Mujahideen: The Enemy Within* (New Delhi: Hachette India, 2011), p. 61 and p. 293.

disruption, the vast bulk of the Indian counterterrorist effort is going into defensive surveillance. This has left the terrorists with the initiative and they are using it to devastating effect.

Threat Assessment for India

Following the killing of Osama bin Laden barely a mile from the Pakistan Military Academy in May 2011, international scrutiny of the ISI has grown more intense. There is considerable scepticism of the idea that the world's most wanted terrorist could live for several years in a garrison town without being detected by local intelligence agencies. However, this does not mean that foreign governments have started to share India's perception of the ISI's relationship with terrorist groups. Publicly at least, the 'rogue faction' thesis remains fashionable since it allows Pakistan scope to amend its behaviour. Any accusations of 'state-sponsored terrorism' would have to be substantiated by evidence linking top Pakistani decision-makers to a specific terrorist incident. Since ISI involvement with *jihadist* groups occurs primarily at sub-departmental levels on an ongoing basis, no such evidence is likely to become available. The Western analytical model will not apply to India's predicament.

In effect, India will remain diplomatically isolated in its efforts to combat cross-border terrorism. With a local support base continuing to exist for *jihadist* groups, New Delhi shall be constrained from exclusively blaming the ISI. Even if such support is limited to a microscopic number of disaffected youth, it shall still provide the ISI with plausible deniability in covert action. Indian police have already revealed themselves to be, at best, partially competent in investigating previous terrorist incidents. Further accusations against Pakistan shall not have much credibility, unless endorsed by third countries as happened after the 2008 Mumbai attack. In this case, foreign governments merely supported India's claim that the attack originated from Pakistan—a fact that had anyway become clear once the identity of the sole captured terrorist, Ajmal Kasab, was confirmed. Despite this, Islamabad initially attempted to portray the carnage as an example of domestic unrest within India and flatly denied any involvement of its nationals.[17] Its behavioural pattern since then does not provide grounds for optimism, considering that Pakistani investigations into the Mumbai conspiracy have stalled.

Jihadist groups are aware of India's isolation. In their assessment, Western statements of support for New Delhi are only intended to blackmail Islamabad: Reminding the ISI that its 'rogue' operations can be publicised, if it does not cooperate with the West.[18] With the agency continuing to help disrupt *jihadist* plots against Western homelands, such statements are unlikely to progress into any kind of follow-up. LeT and its Indian surrogates will continue planning further attacks on India while taking care to avoid leaving a strong evidentiary trail as happened in Mumbai. Their attacks will primarily consist of IED explosions unless the Indian Mujahideen can raise a squad of suicidal gunmen and replicate the massacre/siege situation of 2008. Intelligence agencies need to be especially vigilant against such a scenario as it would massively raise the threshold of domestic subversion.

From developments that have unfolded since 2000, two main trends have become evident.

Firstly, the ISI wishes to have total control over Pakistan-based groups that target India. For this reason, it has chosen the LeT to act as an intermediary with Indian *jihadists*. LeT is an Ahle Hadith (i.e., Wahhabi) group whose ideology resonates more in the Arab Middle East than in South Asia. In jurisprudential terms, its support base within Pakistan is miniscule, compared to the majority of that country's *jihadist* organisations which follow the indigenous

[17]Pakistan took forty-two days just to admit that Ajmal Kasab was one of its citizens, and did so only because there was no point in continuing to deny it anyway.

[18]Shahzad, Saleem, *Inside Al Qaeda and the Taliban* (London: Pluto Press, 2011), p. 94.

Deobandi sect of Islam. The risk of domestic blowback is, therefore, limited. Logically, it would have made more sense to use Deobandi groups to subvert Indian Muslims, who are also predominantly Deobandis. Instead, the ISI has opted for a fragile alliance between LeT and the Students' Islamic Movement of India (SIMI), another Ahle Hadith group. It seems reasonable to assert, from this pattern, that the ISI's main priority is not damage maximisation, but retaining maximum control.

Secondly, for much of the last decade, the ISI has been using Bangladesh and Nepal as transit points for hiding its involvement with the Indian Mujahideen. It has enjoyed support in this endeavour from elements of the Bangladeshi Directorate General of Forces Intelligence (DGFI).[19] During the bombing campaign of 2005-08, Indian investigators frequently blamed the Harkat-ul Jihad Islami (Bangladesh), known as HuJI-B, as being the main culprit. In light of information obtained after the 26/11 attacks, however, it appears that their analysis was wrong. HuJI-B only provided safe houses to LeT operatives in Bangladesh. These operatives in turn, functioned as cut-outs between India-based *jihadists* and the LeT leadership in Pakistan. Interestingly, any travel of Indian Mujahideen members between Bangladesh and Pakistan occurred on Pakistani passports. It is not clear whether these were fraudulently obtained, or were supplied by the ISI. The latter is more likely.

Since 2008, a new line of attack has appeared: The maritime realm. From available reports, it looks as though southern India is now the ISI's priority. Groups such as the Kerala-based Popular Front of India (PFI) are going to be the new face of cross-border terrorism.[20] Relying on sleeper cells implanted within the coastal communities of

---

[19] Nandy, Bibhuti Bhusan, 'DGFI: State Within a State', in Saikia, Jaideep ed., *Bangladesh: Treading the Taliban Trail* (New Delhi: Vision Books, 2006), p. 181.

[20] Lakshman, Kanchan, 'Kerala: Terrorist Entanglements', *South Asia Intelligence Review*, Vol. 7, No. 37, 2009.

Sri Lanka and the Maldives, the ISI-LeT combine will aim to build a supply network of arms and explosives to the southern coastline.[21] The eventual aim would be to carry out *fidayeen* attacks in economic hubs like Bangalore, and possibly, to attack shallow-water shipping. The worst case scenario which India needs to be ready to prevent, and if that fails, retaliate to, is a sea-jacking of an oil platform or high-capacity passenger vessel.

The instinct to strike at India has become integral to ISI operational culture. From available data, it appears that an open competition exists between various LeT operational detachments and their respective ISI handlers to outdo each other in attacking India. The 26/11 attacks, it is believed, occurred partly due to the desire of one ambitious ISI officer to prove that his team could cause more damage in a single strike than a rival team had caused in multiple bombings over the preceding months. The experience of the Soviet-Afghan War suggests that the ISI relishes what one writer has called 'Wile E Coyote-style' attacks on critical targets.[22] Although planning and preparation for such operations continues without interruption, the ISI only implements them under a democratic government. This is because India automatically comes under strong international pressure to avoid undermining democracy in Pakistan by launching a military reprisal. Furthermore, the Pakistan Army is able to outsource the task of diplomatically defending Pakistan to civilian leaders.

[21] In a CLAWS seminar held on 17 December 2013, Lt Gen (Retd) PC Katoch particularly singled out the Maldives as being a likely launching pad for *jihadist* attacks on India. His concerns appear justified to this author. It needs to be stressed that the weak policing capacity of the Maldivian state, combined with simmering political tensions, is creating a festering threat. Just as Al Qaeda used the political space created by the 2011 Arab uprisings to consolidate its presence in the Middle East and North Africa, so too might LeT do the same in Maldives if provided with a suitable opportunity.

[22] Coll, Steve, *Ghost Wars: The Secret History of the CIA, Afghanistan and Bin Laden, from the Soviet Invasion to September 10, 2001* (New York: Penguin, 2004), p. 134.

In the final analysis, India needs to strengthen both its defensive and offensive counterterrorist capabilities. It needs to create a centralised analysis cell for tracking terrorist behaviour. Political subversion acts as a bridge between the ISI's official role (espionage) and its assumed role (supporting terrorism). Such subversion is predominantly a non-violent activity that involves silently injecting an ideological virus into society that turns citizens against the government. It needs to be countered if terrorism is to be contained. Human intelligence networks need to be strengthened at the police station level. This requires additional funds and specialised training for policemen in agent recruitment and handling. Special Weapons and Tactics (SWAT) capabilities have to be raised at the *thana* level in localities where intelligence reports suggest that LeT *fidayeen* could strike.

On the external front, agent penetration operations need to be mounted to compromise the operational security of Pakistani *jihadist* groups, so that key activists can be neutralised and command and control disrupted. Although dialogue with Pakistan can be maintained for the sake of appearances, India should be clear that this is only a sideshow, to provide political cover for covert reprisals. In order to carry out these reprisals, New Delhi will have to clandestinely employ special operations to inflict intolerable pain on individuals directly involved in assisting acts of cross-border terrorism. The next chapter shall show how lack of response to Pakistani provocations during the Soviet-Afghan War emboldened the ISI, prompting it to turn its proxy war mindset eastward, in the direction of India.

6

The Bear's Claws in Afghanistan

This chapter will examine the role played by Soviet special operations forces (known as Spetsnaz) in combating Pakistani proxy warfare during the 1980s. The combat theatre was Afghanistan, which had been occupied by the Soviet Union for the sake of strengthening a puppet regime. Backed by the combined intelligence, financial and material resources of the Western and the Arab/Islamic blocs, Pakistan embarked on a policy of 'fighting the Soviets to the last Afghan'.

The Soviet-Afghan War is still viewed by the Pakistani security establishment as a swashbuckling adventure and a great victory: An example of defeating a superpower to the point of its disintegration.[1] More than any other event, this conflict gave the ISI and Pakistan Army hope that India, too, would break up as a result of attritional warfare, waged through proxies acting on Islamabad's behalf. It is,

[1]Schofield, Carey, *Inside the Pakistan Army* (New Delhi: Pentagon Security International, 2011), pp. 61-64.

therefore, vital to revisit the dynamics of the war, as a means of understanding how it warped Pakistan's strategic psyche.

Even as the war was ongoing, Indian analysts realised that it would eventually expand into J&K, as the ISI would seek to replicate its 'success' against India. Our security policy in J&K has, therefore, been conditioned in part by a need to avoid mistakes made by the Soviets; in particular, relying excessively on air power in an age of man-portable air defence systems (MANPADS). We have also appreciated the importance of denying Pakistan the ability to internationalise its proxy war and obtain the cooperation of other governments in sponsoring Kashmiri separatism.

Need to Defeat Pakistan's Strategy

What India still has not understood is what the Soviet-Afghan War revealed about the Pakistani military's own thought processes, regarding what is good for Pakistan. While most civilian analysts argue that the anti-Soviet *jihad* led to the proliferation of drugs and gun culture, and strengthened the ISI at the expense of civilian intelligence agencies, military thinkers in Pakistan remain fixated on the idea of 'strategic depth', which they have pursued since the War.[2] The suffering inflicted on thousands of Pashtuns on both sides of the 2,640 km-long Durand Line, plus the domestic instability caused to Pakistan, is of no concern to them. Safe in their cantonments and retirement bungalows, they fantasise about controlling access to Central Asia and waging an endless *jihad* against India. Notwithstanding the events of the last fifteen years in remote border areas adjoining Afghanistan, this class of strategists remains influential in Islamabad and Rawalpindi, and it is regaining prominence.

Those in India who believe that attacks by the Pakistani Taliban and sectarian violence will somehow, force a reassessment of the

[2] Hilali, AZ, 'The costs and benefits of the Afghan War for Pakistan', *Contemporary South Asia*, Vol. 11, No. 3 (2002), pp. 299-306.

ISI and Pakistan Army's grand strategy, are deluding themselves. Pakistan as a state is not about to correct itself in coming years, any more than it did during the 1990s when the Americans and Soviets both disengaged from Afghanistan and left Islamabad to pick up the detritus of waging a proxy war. Even then, Pakistani government spokesmen took the same plaintive tone that they do now, insisting that the world owed them a favour for having fought the Soviets. Yet, when one looks closely at what Pakistan asked for (and got), it was actually more of the same—more control over Afghanistan's internal affairs and more Western tolerance for radical Islamism—the same things that had supposedly landed Pakistan in a perilous situation in the first place.

Rather than hope for a course correction by its neighbour, India ought to focus on defeating Pakistan's proxy war strategy by any means. This includes offensive special operations—the one weapon which the Soviets did not use in any systematic way during the 1980s. Should the Pakistan Army find itself forced to choose between downgrading its grandiose expectations of regional influence, or facing an angry public at home for failing to secure the country's borders, it will choose the former. Even if it decides to call India's bluff and escalate the proxy war, that would not differ in any meaningful way from what it has already demonstrated the capacity and willingness to do: Kill Indian civilians.

Soviet Strategy

It has been remarked that Soviet forces in Afghanistan had a good grasp of combat tactics, but had no strategy to guide the use of these tactics and evaluate their efficacy.[3] This is perhaps not a fair argument: The Soviets did indeed have a strategy, and they implemented it better than most historians give them credit for.

[3] Braithwaite, Rodric, *Afgantsy: The Russians in Afghanistan 1979-89* (London: Profile, 2012), p. 224.

Unfortunately, political intrigue and economic difficulties at home intervened to prevent the strategy from being fully carried out, thus leaving Afghanistan open to Pak depredations.

The Soviets were not interested in destroying the Afghan *mujahideen* (the ISI-supported insurgents). Instead, they only wanted to wage a low-level war using an indirect, intelligence-led strategy. This consisted of three parts: Conducting a War of Attrition in the Afghan hinterland, sealing off borders with Pakistan, and penetrating the insurgents with the help of local intelligence partners. A troop ceiling was set at 115,000, hot pursuit into Pakistan was not permitted in order to limit escalatory risks, casualty avoidance on own side was emphasised, and the Afghan government was propped up to take a bigger role in counter-insurgency. However, once Mikhail Gorbachev came into power with his radical ideas of reform, even these limited commitments were judged too troublesome to maintain.[4]

In military terms, Afghanistan might have been an unnecessary war but it was not a military debacle. Policymakers in Moscow had set an informal limit on the monthly casualty rate, which they thought was 'acceptable'—150 per month.[5] Even during the bloodiest phase of the conflict, troop deaths averaged 148 per month. This was certainly high, but it does not suggest intolerable levels of pain. The situation on the ground told a different story from the statistics of a conflict that slowly saw the occupation forces dominating the country, largely thanks to aggressive use of special operations forces.

The Soviet Spetsnaz did an excellent job in area dominance, but were not used for area penetration. That might have been Moscow's one critical blunder of the war. Soviet restraint did not stop Pakistan (or the West) from supporting the *mujahideen*. However, it did force the Soviet Union into a situation where its troops endured

---

4 Cassidy, *Counterinsurgency and the Global War on Terror*, p. 59.

5 Baev, Pavel K, 'How Afghanistan Was Broken: The disaster of the Soviet intervention', *International Studies Review*, Vol. 15, No. 3 (2012), p. 256.

tremendous hardships while fighting an enemy protected by the inviolateness of international borders: An inviolateness that was, of course, entirely one-sided.

AREA DOMINANCE MISSIONS

Spetsnaz troops were mainly used in two types of area dominance roles in Afghanistan: Terrain seizure and supply interdiction.

TERRAIN SEIZURE

The first challenge facing the Spetsnaz was to help regular infantry battalions dominate the terrain in their assigned sectors. For the first years of the war, they were hampered by the differential in mobility between their own men, and those of the units they were supposed to help. Aside from airborne and reconnaissance units, plus the Spetsnaz themselves, the rest of the Soviet occupation army (some 80 per cent of the total strength) was a road-bound force.[6] It lacked light infantry, which was needed to pursue the *mujahideen* across rough ground. The bulk of its troops were mechanised infantry who were dependent on their vehicles for both transport and fire support. At most, they could venture a kilometre off motorable roads, but in the heat of Afghanistan, further movement was impractical. Infantry weapons were designed for firepower, not accuracy, thus requiring that each soldier carry a heavy ammunition load, in addition to his sixteen-kilogram flak jacket. Crew-served weapons such as heavy machine guns and automatic grenade launchers were too heavy to be carried long distances.[7] The mismatch in speed between special operations forces and regular forces, thus, created deployment gaps which the enemy, with his superior local knowledge, invariably exploited to make good his escape.

---

[6] Ghose, R, 'Air Power in Sub-Conventional Warfare in the South Asian Region: Past Lessons and Future Trends', *Air Power*, Vol. 1, No. 5 (2010), p. 9.

[7] Cassidy, *Counterinsurgency and the Global War on Terror*, p. 55.

In 1984, a breakthrough arrived in the form of a successful offensive in the Panjshir Valley. The Soviets threw in 20,000 newly-inducted troops, of whom 10 per cent were Spetsnaz, against *mujahideen* leader Ahmed Shah Masoud. Although Masoud's intelligence network picked up warning signs of the attack and he withdrew to safer areas, the suddenness with which Soviet commandos struck surprised him. Previously, he had been able to retreat in good order in the face of slow-moving Soviet assaults, but the additional troops upset his equipoise. Highly fit and armed with advanced communication equipment, Spetsnaz teams were able to race up mountain slopes as fast as the retreating guerrillas, leading to a breakdown of command and control. Using helicopters, the Soviets landed blocking forces astride the *mujahideen*'s escape paths, inflicting heavy casualties. Meanwhile, armoured troops advanced along the main roads, pulverising any resistance that they encountered.[8]

Over the following years, the task of area dominance became easier, as the *mujahideen* grew wary of remaining affixed to a stronghold and thereby presenting a positional target for Soviet forces to concentrate on and destroy. Using 'hammer and anvil' tactics, the Soviets developed a simple way of dislodging the guerrillas from any locality: Infantry and armoured columns would mount a frontal assault, while Spetsnaz troops would be heli-dropped to set up ambushes along likely escape routes.[9] Lightly armed and operating in small teams, the commandos moved stealthily across broken terrain and set up observation posts at vantage points. They were, thus, able to guide the regular forces as they chased the fleeing guerrillas. Tactical mobility and fire support from helicopters made the Spetsnaz a deadly foe, whose numerical weakness was more than made up for in fighting ability.

---

[8] Coll, *Ghost Wars*, p. 122.

[9] Burli, RG, 'Offensive Air Power in the high Mountains', *Air Power*, Vol. 2, No. 1 (2007), pp. 89-90.

SUPPLY INTERDICTION

Large-scale offensives were a rarity; the bulk of Spetsnaz activity consisted of interdicting *mujahideen* supply caravans that were entering from Pakistan. In 1984 the Soviets began Operation Curtain (renamed three years later as Operation Barrier). The aim was to choke off *mujahideen* supplies by constant patrolling near the borders and laying ambushes. On average, at least one ambush would be carried out every night. Typically, intelligence analysis would identify the best place and time to lay an ambush. Commandos would then be helicoptered to the vicinity, being inserted anywhere between ten and thirty kilometres from the actual spot to throw off *mujahideen* spies. They would then approach the site by foot, and usually under the cover of darkness. Often, several decoy landings would be made by the helicopters, while the troops disembarked only at one spot along the way.[10]

A Spetsnaz ambush would follow a standard pattern: The first vehicle in a *mujahideen* convoy would be allowed to pass since it was usually scouting ahead of the main column. The second vehicle would be stopped by a silenced weapon, so as to cause the vehicles behind to bunch up without suspecting anything was amiss. The Spetsnaz team leader would then trigger the ambush either by detonating mines emplaced along the road, or by firing from a sniper rifle. All weapons would be emptied at the convoy on full automatic for the first magazine, in order to create a shock effect. Thereafter, each of the ambushers would fire freely within their assigned sectors. A typical mission would be between three and four days long. The team would carry about forty-five kilograms of equipment and ammunition per man. Most ambushes were carried out by platoon- or company-sized forces, operating within a 200-kilometre radius

---

[10] Doohovskoy, Andrei A, *Soviet Counterinsurgency in the Soviet Afghan War Revisited: Analyzing the Effective Aspects of the Counterinsurgency Effort*, MA Thesis presented at Harvard University in September 2009, http://boekje-pienter.eu/images/coin-soviet-afghanwar.pdf, p. 97.

of their main base, which would be adjacent to an airfield.[11] There were seven such bases across Afghanistan in the early 1980s, before the troop surge of 1984, with each base housing 250 commandos, making for a total of 1,750 Spetsnaz troops.

To locate caravans, the Spetsnaz used satellite imagery, infrared detectors and communications interception vehicles, colloquially called 'Omsk vans'.[12] These vehicles would listen out for radio traffic from convoys approaching the Afghan border on the Pakistani side. They were aided by poor communications security: *Mujahideen* commanders tended to let their guard down when in Pakistan, believing that they were still on safe ground. Sometimes, the Spetsnaz would send in MI-24D Hind gunships to strafe the convoys when they were still five to ten kilometres inside Pakistani territory. Caught in the open and tactically unprepared to deal with an ambush, the guerrillas would be decimated, even as Soviet commandos heli-landed around them to conduct mopping up operations.[13]

Human intelligence was a bigger problem: Spetsnaz troops had no training in Afghan languages. They did not have much contact with the local population and thus, could not learn through personal interaction. Whatever human intelligence was gained, came through agents run by the GRU (Soviet Military Intelligence). These agents were often former *mujahideen* who had defected to the Afghan government and had been loaned to work with the Soviets. They provided an entry point into the guerrillas' secret world. Using them as frontmen, disguised Spetsnaz commandos were able to employ pseudo-terrorist tactics: Operating with blackened faces at night, they masqueraded as *mujahideen*, seeking out guerrilla leaders in

[11] Kinnunen, Eero and Grau, Lester W, 'Two Tours in Afghanistan: Twenty Years and Two Armies Apart', *Military Review* (May-June 2011), p. 42 and p. 45.

[12] O'Ballance, Edgar, *Afghan Wars: Battles in a Hostile Land, 1839 to the Present* (London: Brassey's, 2002), p. 155.

[13] Coll, *Ghost Wars*, p. 134.

their safe havens and neutralising them.[14] Unfortunately, this was just about the only measure that the Soviets took, to hit the guerrillas where it really mattered—at home.

(LACK OF) AREA PENETRATION MISSIONS

The Soviets made a strategic blunder by not attacking *mujahideen* sanctuaries in Pakistan, for fear of escalating the war and potentially drawing in the US. Given the massive support that the ISI, CIA and Saudi intelligence were providing the *mujahideen*, in hindsight it would have made more sense to inflict severe costs on Pakistan, the most proximate of the insurgency's sponsors. As events turned out, Soviet restraint only emboldened Pakistan to push in ISI operatives and Special Services Group (SSG) troops to aid the *mujahideen* directly.[15] Approximately 25,000 Pakistan army soldiers saw action against the Soviets while operating under the flimsiest of civilian cover. Over a period of time, this cross-border involvement allowed Pakistan to claim a seat at the negotiating table, when the Soviets began to send out feelers for a political settlement to the conflict. The significance of this achievement should not be understated: A country that had insisted all along it was not interfering in the affairs of a neighbouring sovereign state, was able to cash in on past (as well as ongoing) transgression and become a stakeholder in determining the future of that state.

THE KRER RAID: AN EXCEPTION THAT PROVED THE RULE

Whatever cross-border raids occurred, took place in defiance of established Soviet policy, and were the result of restlessness among local Spetsnaz commanders. The most famous of these raids was launched in March 1986. A year previously, the *mujahideen* had

---

[14] [Anonymous], 'The Soviet Elite: Spetnaz', in Lewis, Jon E ed., *SAS and Elite Forces* (Bristol: Parragon, 1996), p. 50.

[15] Mir, Amir, *Talibanization of Pakistan: From 9/11 to 26/11* (New Delhi: Pentagon Security International, 2009), p. 12.

destroyed a Spetsnaz company, and the dead men's comrades wanted revenge. They knew that the *mujahideen* unit which carried out the attack was based at Krer, a location just across the border from Kunar province (about one kilometre inside Pakistan). After capturing an insurgent who lived at the base, they learnt important tactical details about its strength, weaponry and defensive protocols. A reconnaissance raid was launched to test the accuracy of the information, and seize further material of intelligence value.

Based on the information gathered from the prisoner and the reconnaissance raid, a two-battalion assault was launched on the night of March 30. It had little chance of success because the base was heavily fortified and the occupants were aware that a probing attack had just been conducted. The only reason that the operation did not turn into a complete disaster was accidental: Instead of advancing along their planned route, the commandos lost their way in the darkness (despite having a guide) and inadvertently outflanked the main *mujahideen* defences. The defenders had been expecting the Spetsnaz and had arrayed their weaponry along the most likely route of advance. They were caught off guard by the Soviets' mistake. Even so, they counterattacked in strength at daybreak.

The commandos were slowly ground down by superior enemy numbers until helicopter gunships arrived. After first refusing to provide fire support, despite repeated requests, they cleared the *mujahideen* from the immediate area and allowed the two battalions, which had been separated during the fight, to regroup. The helicopter pilots later explained that they had deliberately not attacked for thirty minutes, so that they could deny ever having fired across the border. Since they expected their superior officers to question on whose authority they had opened fire, they ensured that the cockpit tapes recorded the repeated Spetsnaz requests and the pilots' denials. After half an hour of this, they had turned off the recorders and gone in for the attack. Overall, the Spetsnaz raid was an impressive feat, since it took on an estimated 400-500 *mujahideen* in a well-defended position. However, the brashness of the raid meant that it was not coordinated with air power until heavy casualties had been incurred.[16]

Unlike most special operations, where the enemy is caught unawares and one's own side is prepared to attack, the operation at Krer reversed standard military logic. Instead, the enemy was forewarned that an attack was coming, and the Soviets' top commanders were kept in the dark, to prevent them from interfering with the assault. This episode brings out the need for a whole-of-government approach to cross-border special operations: It is vital to authorise raids at the highest political level, so that commanders in the field can be confident about receiving real-time assistance if they run into tactical difficulties during an attack. Whenever the Spetsnaz faced stiff resistance from the *mujahideen*, on either side of the Afghan-Pak border, they had to urgently call for air support. Its absence was literally a death sentence for them, as happened on some occasions.

OVER-DEPENDENCE ON AIR POWER

The war in Afghanistan was going well for the Soviets, after five years of hard fighting, when in April 1986, the CIA provided Stinger missiles to the *mujahideen*, via the ISI. This one development changed the nature of the conflict. Until that time, the Soviets had controlled Afghan airspace completely. Their real strength was a unique type of helicopter, which had no counterpart in the West: The MI-24D Hind. It was both a transport helicopter and a gunship, which made it ideal for special operations. Each aircraft could transport eight fully-equipped troops, and was armed with 80 millimetre rockets and

---

[16] Grau, Lester W and Jalali, Ali Ahmad, 'Forbidden Cross-border Vendetta: Spetsnaz Strike into Pakistan During the Soviet-Afghan War', *Journal of Slavic Military Studies*, Vol. 18, no. 4 (2005), pp. 661-672.

a twenty-three millimetre cannon. Heavily armoured against impact from even high-calibre machine gun fire, it was almost indestructible by infantry weapons. With a speed of 275 kilometres per hour, it could transport both men and firepower anywhere within a 300-kilometre radius.

Between them, the Spetsnaz on the ground and the Hinds in the air exerted relentless pressure on the *mujahideen*, whose morale was plummeting by late 1985. Worried that the Soviets might actually be winning the war, the CIA gave in to persistent lobbying by the ISI, and supplied Stinger missiles. The move risked escalating the war, since it proved beyond any doubt that the West and Pakistan were supporting the *mujahideen* by supplying them advanced weapon systems. There was even concern in the CIA that the Soviet Union might invade Pakistan in retaliation. However, as it turned out, this worry was overblown; the Soviets were shocked and dismayed, but lacked courage for expanding the war further. The *mujahideen* regained their fighting spirit once the Soviets lost control of the air.

Widespread availability of Stingers completely changed the ground-level dynamics of the war.[17] Casualty evacuation by air became difficult, leading to risk aversion among ground commanders, who did not want to put their troops in harm's way. Close air support was no longer possible because all aircraft had to stay above 12,500 feet to remain outside the missile's effective range. Since ground

[17] Not only were the Soviets forced to limit the tactical use of airpower, but the introduction of the Stinger also emboldened the ISI to begin launching attacks against the Soviet Union itself. Teams of Afghan saboteurs, acting on Pakistan orders, attacked military and economic targets up to a shallow depth inside the southern Soviet republics. Between December 1986 and April 1987, these operations increased in scale, until the Soviets issued a stern warning following a particularly provocative attack, that they would respond militarily to further provocation. Upon CIA urging, as well as the nervousness of his own superiors, the head of the ISI's Afghan Bureau, Brigadier Mohammad Yousaf, was compelled to cease the covert offensive. Harclerode, Peter, *Fighting Dirty: The Inside Story of Covert Operations from Ho Chi Minh to Osama Bin Laden* (London: Cassell, 2001), pp. 563-566.

units (regular infantry as well as Spetsnaz), had grown tactically and psychologically dependent on air cover to close in with the enemy, the absence of an aerial shield was debilitating.

All told, 200 Hinds and 133 other rotary wing aircraft were shot down by the *mujahideen* between 1986 and 1989, besides 118 fixed wing aircraft.[18] Coming on the back of difficult slogging matches fought during the first half of the decade, these equipment losses demoralised the Soviets and led them to conclude that Afghanistan was an unwinnable war. Even if it was possible to regain the strategic initiative, most policymakers in Moscow did not feel the sacrifice would be worth the returns. Bereft of air cover the Spetsnaz, which until then had been used to proactively hunt down guerrillas, were reduced to the role of elite infantry—still tougher than other units, but no longer deadlier.

FIGHTING DEFENSIVELY, LOSING THE INITIATIVE

Throughout the war, most Spetsnaz operations were focused on securing the borders. Yet, recent estimates suggest that perhaps only 20 per cent of all *mujahideen* supplies were interdicted.[19] The Afghan-Pak border was too long and too remote to be effectively policed—a situation which favoured the offensive. By not using their special operations forces aggressively, the Soviets ensured that even though they dominated the terrain inside Afghanistan, the geopolitical dynamics of the conflict could be controlled by Pakistan. In total, 657 officers and men from the Spetsnaz died in the war—a toll that, if it had been incurred during area penetration missions, would have logically been accompanied by severe damage caused to Pakistan and its ability to continue supporting the *mujahideen*.[20]

[18] Cassidy, *Counterinsurgency and the Global War on Terror*, p. 56.

[19] Braithwaite, *Afgantsy*, pp. 133-134.

[20] Plater-Zyberk, Henry, *Russia's Special Forces*, Conflict Studies Research Centre, September 2005, p. 5, http://studies.agentura.ru/centres/csrc/rsp.pdf, 24 October 2013.

Spetsnaz troops deployed in Afghanistan were quite poorly equipped when compared to Western special operations forces. Their standard weapon was the AK-74 rifle. There was an acute paucity of silenced weaponry, and whatever was provided (in the form of two AKM rifles with PBS suppressors), was useless because subsonic ammunition was difficult to obtain and because the silencers were destroyed by the muzzle pressure of regular supersonic ammunition. Only unit commanding officers and 2/ICs had night vision binoculars, and only one in ten SVD sniper rifles had a night scope. The main priority of the Spetsnaz supply directorate was to issue commandos with rugged, low-maintenance weaponry rather than fragile high-quality weapons.[21] Despite these handicaps, Soviet commandos almost won the war in Afghanistan, by harassing the *mujahideen* through constant attacks supported by air power. Had it not been for the American decision to induct Stinger missiles into the conflict, plus Gorbachev's own domestic preoccupations, it is possible that the Soviet Union would have prevailed over Pakistan and the West in their proxy war.

To a large extent, the Pakistanis were plain lucky in Afghanistan. This explains why they bungled the Jalalabad offensive of 1988, having failed to read the military situation correctly. Hoping for quick returns, the ISI's Afghan bureau constantly looked for strategic weaknesses to hit, such as logistical chokepoints. It was obsessed with closing the Salang Tunnel, which connected northern and southern Afghanistan, and through which 75 per cent of all Soviet road traffic passed. The bureau came up with a scheme to simultaneously detonate three truck bombs in the tunnel during winter, causing it to collapse and thereby wreaking havoc upon the Soviet supply situation. Had the plan succeeded, it may have indeed had a strategic impact. However, in an example of the ISI's imperfections, the

---

[21] Kenner (pseudonym), 'Soviet Army Spetsnaz', pp. 98-99.

agency found that it could not rely on local proxies to do the job.[22] In-country advisory teams had little hold over the *mujahideen* leaders once they were operating on their own turf. From being the controlling authority in the *mujahideen* camps (based in Pakistan) they were mere 'guests' in the conflict zone. Other than practice the litany of combat techniques which they had learnt from CIA trainers, ISI operatives in Afghanistan actually did little to influence Soviet decision-making over the costs and benefits of the war.[23] Had Moscow decided to escalate by taking the conflict back into Pakistan itself, there is little the ISI could have done about it, except run to the CIA for help.

Pakistani Hubris, Post 1989

The Pakistanis choose to believe that the brilliance of their generals and the intrepidness of their spies drove the Soviets from Afghanistan. This delusion has played no small role in shaping the ISI's strategy towards J&K since the early 1990s. For a long time the ISI assumed that once India was forced to begin negotiating on the status of Kashmir, as a result of an Afghanistan-style slow bleed of its military forces, international mediation would step in. The Pakistanis have blithely thought that the end product of such mediation would be a settlement favourable to themselves, and one which India would be forced to accept due to hostile international opinion and lack of military alternatives.[24] Their aim is, therefore, to replay the Afghan

---

[22] Yousaf, Mohammad and Adkin, Mark, *Afghanistan the Bear Trap: The Defeat of a Superpower* (Barnsley, South Yorkshire: Leo Cooper, 1992), pp. 74-76.

[23] CIA trainers taught ISI operatives a total of 60 combat techniques. These included how to build improvised explosive devices (IEDs), neutralise sentries, handle agents in high-risk zones, and use armour piercing ammunition against strategic targets (a skill that the ISI would seek to employ against India during the Khalistan militancy). Cooley, John K, *Unholy Wars: Afghanistan, America and International Terrorism* (New Delhi: Penguin, 2001), pp. 88-91.

[24] Haqqani, Husain, *Pakistan: Between Mosque and Military* (Washington DC: Brookings Institution Press, 2005), p. 289.

jihad step-by-step and cause death by a thousand cuts to India, just as they thought had happened to the Soviet Union. In their optimism, they have overlooked the fact that while the Soviet Union was fighting in a foreign land, and its army consisted of conscripts, the Indian Army is fighting on its own territory, and is a tough, professional force.[25] Also, it has no shortage of manpower, or of experience in counter-insurgency.

Until the global shock generated by the 9/11 attacks in the US, Pakistan's aim was to defeat India using a Vietnam/Afghanistan model. Either the Indians would just quit fighting from sheer frustration, as the Americans did in Vietnam, or they would be compelled to by hostile international reaction, as the Soviets did in Afghanistan. To the Pakistani security establishment's own frustration however, neither scenario materialised. India managed to avoid being stigmatised as a repressive state, and instead, it was Pakistan that in 1993 came close to being labelled a state sponsor of terrorism. Even as a lack of strategic imagination kept the ISI and Pakistan Army on their pre-set course, India has managed to improve its relations with the US and emerged as a key partner of the West.

Credit for this turnaround should go to the Indian political and military leadership of the 1990s, who understood the country's limited options and crafted a strategy accordingly. The Indian Army adapted splendidly to the counter-insurgency task by setting up the Rashtriya Rifles (RR) in 1990 as a force dedicated to maintaining internal security in J&K. The RR's formation helped avoid many of the negative side effects that counter-insurgency has upon war-readiness. It has provided a platform for battle inoculation of officers and men, for developing dual competence in urban and rural

[25] Subramanian, LN, 'CI Operations in Jammu and Kashmir', *Bharat-Raskhak Monitor*, Vol. 3, no. 2 (2000), http://www.bharat-rakshak.com/MONITOR/ISSUE3-2/lns.html, 24 October 2013.

operations, and for experimenting with small-unit tactics. Each RR battalion has six rifle companies as opposed to the four commonly found in an infantry battalion, thus allowing for greater operational flexibility. By focusing on internal security duties, the force has freed up regular infantry battalions to guard the borders.[26] India, thus, does not have to make a tradeoff between having a first-class army and adapting for irregular warfare, in the way that the Americans and the Russians were.

From Proxy War to Proxy Blackmail

After 9/11, a slow realisation seems to have dawned upon Pakistan that it could no longer hope for residual ideological hostility left over from the Cold War era, to get the West to line up against India. It has, therefore, switched to a new strategy, one which we are still struggling to find an answer to: Indirectly internationalising the J&K proxy war through transnational terrorism. Pakistan is leveraging its unique position as a safe haven for global *jihadism*, by rationing out counterterrorist cooperation to countries that support its stance on J&K. Thus, the UK, with its large and unassimilated Pakistani minority, has begun to toe Islamabad's line in exchange for help in suppressing *jihadism* at home.

The UK has played a key role since 26/11 in covering up Pakistani sponsorship of anti-Indian terrorism, even going so far as to act as Pakistan's advocate to the US and the European Union. Perhaps this is understandable: With 400,000 British-Pakistanis travelling back and forth between the two countries each year, British intelligence simply cannot do without the ISI's help in tracking *jihadist* networks. However, the UK's tendency to gloss over Pakistani terrorism directed at India is a worrying sign that the ISI has found a new, more

---

[26] Chatterjee, Debalina, 'Rashtriya Rifles in Kashmir: India's Counter-terrorist Force', *Air Power*, Vol. 7, No. 1 (2012), p. 98 and p. 106.

dangerous strategy: If it cannot coerce India directly, it can get other countries to do so by allowing its citizens to pose a terrorist threat to their homelands. In effect, what Pakistan is trying is to replicate the successful Arab effort to internationalise the Israel-Palestine conflict, by covertly promoting international terrorism.

7

Israeli SOF and Counterterrorism

This chapter makes a reverse argument to the preceding one: While the Soviet-Afghan War showed that lack of area penetration can embolden irregular adversaries, the Arab-Israeli conflict cautions us that overuse of area penetration can also have negative consequences. It can end up internationalising a regional dispute, by prompting neutral countries far removed from the complexities of the situation to bandwagon against the stronger side. Irrespective of the merits of its case, a country that is perceived (rightly or wrongly) to readily use force leaves itself vulnerable to a creeping international isolation. Thus, even if a state-sponsor of terrorism is made to suffer militarily, it can still walk away with a political victory if the victim state is not as proactive about shaping international opinion in its own favour, as it is in launching reprisals. The Israelis, believing that they face an existentialist threat and must fight for their survival, have done a bad job of strategic communication when compared with their Arab adversaries. As one author has noted, the operational reach of Israel's intelligence services has exceeded their strategic grasp of geopolitical factors.[1] India must avoid falling into this trap, if it uses special operations to inflict pain upon Pakistan-based terrorist groups.

'War of Attrition' and Transnational Terrorism

Israel's experience of combating irregular warfare has two components that are relevant to India. The first, and less well-known, is the 'War of Attrition' waged by Egypt during 1968-70. This conflict was characterised by cross-border shelling, commando raids, and air strikes. Readers would immediately recognise similarities with the current situation on the Line of Control (LoC). Notwithstanding the so-called 'ceasefire', Pakistan is hell-bent on disturbing the hard-won tranquillity within J&K, by engineering clashes on the LoC and then seeking international mediation. This is what Egypt tried to do as well after it had lost the Six Day War of 1967 and in the process, also lost the Sinai Desert to Israel. Determined to stop Israel from consolidating its political administration over the Sinai, Egypt resorted to an attritional strategy, hoping to convince Israel that the cost of holding on to the conquered territory would be too great. Simultaneously it sought to draw in the Soviet Union on its own side, much as Pakistan today wishes to bring in China. Although India remains legitimately on its own territory, and instead, it is Pakistan which is illegally occupying a large part of J&K, these nuances have not stopped the Pakistani army from trying the same methods previously used by Egypt.

The second component of Israel's counterterrorist experience which is relevant to India is transnational terrorism by the Palestinian Liberation Organization (PLO) and Hizballah. Both these

---

[1] Jones, Clive, 'A Reach Greater than the Grasp': Israeli Intelligence and the Conflict in South Lebanon 1990–2000', *Intelligence and National Security*, Vol. 16, No. 3 (2001), p. 1.

organisations have understood that Israel can be isolated from its supporters in the West through a game of 'divide and conquer'. By targeting Israeli nationals in horrendous attacks that have an international impact and a logistical footprint across many countries, the PLO and Hizballah have signalled that despite being non-state actors, they can do serious harm to state adversaries. So long as neutral countries do not interfere with their activities, and do not otherwise tilt in favour of Israel, their citizens will be left alone. Although this is nothing more than blackmail, in the world of realpolitik it still works in forcing governments to prioritise their own security interests above those of their allies. Pakistan has learnt this and is doing the same through LeT: Threatening the West indirectly in order to get Western countries to toe its line on J&K, or at the least, not criticise its sponsorship of terrorism against India, a democracy and natural ally of the West.

Need to Appreciate the Limitations of Special Operations

Israel's special operations forces are justifiably renowned for their daring and tactical sophistication. Yet, what is not well understood about them is how their ability to shape strategic dynamics has been circumscribed by the enemy's adaptability. Egypt, the PLO and Hizballah all learned to fear the capabilities of Israeli SOF, but they were not deterred by these. Instead, they merely factored in the damage that could be caused to their activities, and carried on regardless. India cannot allow itself to become predictable like Israel; it cannot allow Pakistan the prestige of standing up to a powerful neighbour despite heavy punishment. If and when India chooses to retaliate to Pakistan-sponsored terrorism, it needs to do so in a manner that synergises political and operational considerations, rather than allowing these to diverge. Only then will India weaken Pakistan's will to wage proxy warfare.

Special Operations in the War of Attrition

Neither Israel nor Egypt won the War of Attrition, but Egypt ended up stronger than it had been before the onset of hostilities. This was not due to good planning: Indeed, both sides were inept at controlling the situation and turning it in their favour. However, the Egyptians benefited from a greater understanding of their own limitations while the Israelis, flush with success after the 1967 victory, found themselves unable to cope with the morale drop that a slowly accumulating body count created.

The war began in September 1968, when Egypt started shelling Israeli positions along the Suez Canal (at the time, the unofficial border between the two countries). Artillery barrages during the day were followed by commando raids during the night. The Egyptians used Ranger battalions—raiding units that operated under divisional commanders—to attack the Israeli outposts. Israeli intelligence was unable to predict when and where the next raid would occur, since most of its human sources were concentrated in higher headquarters, away from the combat theatre.

Every 'Shock' Has a Limited Shelf Life...

After determining that the increase in Egyptian assertiveness was part of a larger trend, and was not just a spike in adventurism by local commanders, Israel retaliated. On the night of 31 October 1968, it carried out a coordinated heli-borne raid against three infrastructure targets roughly 125 miles inside Egypt. Two bridges across the Nile and a hydroelectric station that supplied power to Cairo were destroyed. The attack was aptly code-named Operation Shock, and that was literally what it meant to do: Shock the Egyptians into passivity by making them realise how vulnerable they were.

In the immediate term, the operation was successful. Egypt promptly stopped harassing Israeli defences along Suez Canal and concentrated instead on reviewing its domestic security system.

A new militia was raised, called the Popular Defence Force, to provide a first response in the event of further Israeli raids. Rear area units were instructed on how to coordinate their reactions better. Once these measures were judged to be effective, Egypt resumed attacks on Israeli troops in March 1969.

As fighting dragged on, the Israeli political and military leadership faced a paradox: They wanted a return to normalcy along the Suez Canal, but to achieve this, needed to escalate the level of violence to a level that Egypt could not match. This in effect meant that the situation along Suez Canal would have to get worse before it got better. The general belief was that more area penetration operations, carried out on a reciprocal basis in response to Egyptian raids, would send a signal to Cairo.

Between April and December 1969, Israeli SOF personnel carried out a number of tactically sophisticated raids. In one of these, commandos from Sayaret Matkal and Shayetet 13 (the most elite army and naval SOF units) seized an enemy radar station and created a blind spot in the Egyptian air defence system. Over 100 fighter-bombers poured in through the gap, in a coordinated airstrike, to hit Egyptian positions along the Canal. On another occasion, Israeli commandos disguised themselves as an Egyptian armoured unit, complete with four Soviet-manufactured tanks and three armoured personnel carriers. After clandestinely crossing the Canal, they went on a ten-hour rampage, utterly destroying Egyptian positions over a thirty-mile stretch of roadway adjoining the Canal. They encountered little resistance, as the Egyptians had never thought the Israelis were capable of carrying out an armoured crossing, and so had only prepared their troops for small arms battle.[2]

Notwithstanding these successes, by the end of the year, things were not going so well at the strategic level. The Israelis had hoped

[2] Terrill, W Andrew, 'The nature and value of commando operations during the Egyptian-Israel war of attrition', *Small Wars & Insurgencies*, Vol. 8, No. 2 (1997), pp. 19-28.

that repeated humiliations by their more adept SOF would force the Egyptians to de-escalate. Instead, what happened was that the Egyptians were encouraged by Israeli attacks, seeing them as a sign of desperation. They chose to think that their attritional strategy was working, and that Israeli morale was slowly being ground down by continued fighting.

They were right. Israel had initially begun its SOF campaign by attacking infrastructure targets, but this did not have the desired effect of deterring Egypt. It then switched to attacking military installations, hoping that the ignominy of being unable to protect its own personnel would force the Egyptian military establishment to back down or risk scorn from the civilian populace. These measures, too, did not work, and their changing nature confirmed Egypt's belief that Israel was groping for a solution. All that Egypt needed to do was keep the pressure on, regardless of its own losses.

Israel Over-Reaches Itself

By January 1970, it was clear that special operations had failed to alter Egypt's cost-benefit calculus. The early Israeli commando raids, from April to July 1969, had not been provided with air cover in order to avoid escalating the conflict. After July, the Israeli air force regularly flew sorties in close support of special operations. However, even these did not lead to a lowering of violence. So, in January the Israeli leadership decided to conduct strategic bombing of the Egyptian hinterland.

Egypt panicked. President Gamal Abdel Nasser rushed to the Soviet Union and pleaded for assistance with building an air defence system that could protect the hinterland. Although the Soviets were not keen to embroil themselves in a conflict with Israel, an American ally, Nasser's desperate situation caused them to swing around in favour of intervention. The decisive moment allegedly came when he pointed out that failure to protect Egypt would diminish Soviet Union's own standing in the world.

The Israelis had considered the possibility of Soviet intervention before they had launched their strategic bombing campaign. Yet, in an example of the over-optimism that clouds military thinking during an escalating crisis, they had done a skewed cost-benefit analysis; one that heavily favoured carrying out the bombing campaign. They overestimated the probability that expected benefits from the campaign would actually materialise and did not fully think through the implications of the accompanying costs. They expected that even if the Soviets did intervene, all that would change would be Egypt's ability to protect the airspace over its hinterland. The border zone near the Suez Canal, the area that Israel really wanted to control, would still be vulnerable to Israeli air power.[3]

In fact, Soviet support went a lot further than Israel had bargained for. The Soviets dispatched fighter pilots to fly Egyptian warplanes. Upon realising that they were going to enter into direct hostilities with Soviet servicemen, the Israelis suspended their long-distance air strikes. The US, Israel's superpower patron, took a decidedly cool stance towards Israeli pleas for support. In Washington's view, Israel had unnecessarily escalated its fight with Egypt to a point where Soviet involvement had been precipitated. The prospective blame for any further escalation, thus, fell on Israel, which found that by bombing the Egyptian hinterland, it had gone beyond the limits of American support, and into a collision course with the Soviet Union. It had to climb down.

Relevance To the Current Situation Along the Indo-Pak LoC

To be fair, as far as the India-Pakistan situation is concerned, one should not overemphasise the parallels with the War of Attrition. In 1968-69, Israel was flush with the spirit of victory, and the inevitable sobering of the public mood that was bound to occur,

[3]Shlaim, Avi and Tanter, Raymond, 'Decision Process, Choice, and Consequences: Israel's Deep-Penetration Bombing in Egypt, 1970', *World Politics*, Vol. 30, No. 4 (1978), p. 495.

was fully exploited by Egypt through a well-timed attritional strategy. The Israeli economy was in dire straits by 1970, for reasons entirely unconnected with the situation at the Suez Canal. Thus, a combination of bad news on the military and economic fronts led to an overall sense of defeatism.[4] This is presently not the case with India, which is justifiably angry with continued Pakistani transgression in the face of numerous peace overtures. Although the Indian economy has slowed down, this has not translated into war-weariness, because by not retaliating to the 26/11 attacks, India has already walked the extra mile needed to reduce bilateral tensions. If Pakistan lacks the sense to reciprocate, India cannot do more than it has already done.

What is relevant to the India-Pakistan context, as far as the events of 1968-70 are concerned, is the need to identify the larger role that special operations can play in deterring an enemy. Israel had hoped that its commando raids on infrastructure and military targets would serve three purposes:

- Inflicting retributive pain on the Egyptian state for its provocations at the Canal,
- Compelling Egypt to suspend its artillery and commando offensive, and
- Minimising the probability of further escalation, by signalling to Egypt that it too was vulnerable to being attacked if it ratcheted up tensions further.

The problem was Egypt was less casualty-sensitive than Israel. Although it suffered heavy losses of men and equipment through Israeli attacks (both ground and air), it had the political capacity to absorb these. After the defeat of 1967, the Egyptian government was able to rally its population by projecting Israel as a ruthless and powerful enemy, against whom they needed to unite as heroic underdogs. This went against the implicit logic of the Israeli

[4] Khalidi, Ahmed S, 'The War of Attrition', *Journal of Palestine Studies*, Vol. 3, No. 1 (1973), p. 84.

commando raids, which were meant to provoke a civil–military divide within Egypt by exposing the military's inability to protect Egyptian sovereignty. Basically, Israeli SOF hit at Egypt's exploitable weaknesses too frequently for their actions to have any deterrent effect, because the Egyptian state and society already knew that they were vulnerable. Cairo's real strength—its ties with Moscow—was not affected by commando strikes.

When special operations failed to dampen Egypt's zest for a fight, the only alternative was to raise the level of conflict to a point that Egypt could not match. The strategic bombing campaign was intended to do this. After September 1969, when Israel received its first consignment of F-4 Phantom fighters from the US, the military balance turned hugely in its favour. The Phantom was the most sophisticated fighter-bomber of its time, and was much less vulnerable to Egyptian air defences than all other aircraft in the Israeli inventory. Possession of it gave Israeli military leaders a (false) sense of invulnerability, and they endorsed the strategic bombing campaign almost immediately. In doing so, however, they ended up *reinforcing* and not *reversing* Egypt's strength, because the campaign triggered the Soviet Union to shift out of its past apathy and take an active role in the conflict.

India needs to take note of this precedent: When Pakistani strengths remain unaffected by our commando raids, the raids will in themselves deliver no strategic results. Only if they are worked into a broader plan to isolate the enemy from third-party support, and convert his traditional assets into liabilities or at least, into politico-military dead weight, will special operations deter Pakistan.

Special Operations as a Response to Cross-Border Skirmishes

India needs to answer four questions, as and when it contemplates carrying out special operations in response to Pakistani aggression along the LoC.

- Can retaliation be carried out at an acceptable price?
- What purpose would be served by retaliating?
- How will our friends react?
- How will the enemy's friends react?

The answers will vary, depending on contextual factors—the military balance, US-Pakistani relations, Sino-Pakistani relations, domestic politics within both India and Pakistan etc. However, worry about adverse international reaction should not lead to pre-emptive closure of any option.

By displaying tremendous restraint after the unacceptable provocation of 26/11, India has earned considerable political credit internationally. It can slowly encash this by conducting special operations on a deniable basis inside Pakistan Occupied Kashmir (PoK). Whenever Pakistani troops raid Indian Army positions, Indian commandos can hit targets situated some distance back from the LoC. Simultaneously, 'black' or unattributable psychological operations can be launched to project these attacks as being the work of Al Qaeda splinter cells or local franchises of the Pakistani Taliban.

To foreign intelligence agencies who could track the actual sequence of ground events through signals interception, Indian officials can privately admit that if Pakistan continues raiding Indian border posts, it shall slowly but steadily be made to pay on the LoC, a blood price for what it did in Mumbai in 2008. While we could not kill Pakistani civilians on the streets of Karachi after 26/11, we can kill Pakistani soldiers in PoK. As long as we establish a politico-psychological linkage between the Mumbai attack and our refusal to passively absorb casualties along the LoC, it is unlikely that foreign governments will force the issue, for fear of revealing themselves as hypocrites in counterterrorism.

The big mistake that Israel made in the War of Attrition was in failing to communicate the rationale of its reprisals to the larger international community. It did not shape political discourse over the conflict, thus allowing great power agendas to dictate how its actions were interpreted. The US was not interested in getting into a confrontation with the Soviet Union, and leveraged its unparalleled dominance of the global media to signal that it was not overly sympathetic to Israel. Learning from Israel's experience, India has to ensure that foreign (and particularly Western) views of Pakistan are slanted in favour of doing nothing whenever Pakistani soldiers are killed by unidentified raiders inside PoK. Over time, this would push Pakistan into isolation, if it persists with its tomfoolery against India.

PLO and Hizballah: The ISI's New Role Models?

There is a pattern in some aspects of Pakistani strategic behaviour which suggests the ISI is trying to copy methods developed by the PLO and Hizballah, to influence international opinion. The PLO, (renamed as the Palestinian Authority (PA) after the 1993 Oslo Accords) used international terrorism to attract attention to the plight of Arabs who had been displaced by the creation of Israel. Throughout the 1970s, the PLO either directly planned or indirectly endorsed terrorist acts, including the massacre of Israeli athletes at the 1972 Munich Olympics. Yet, it managed to gain enough political respectability to be widely recognised as the authoritative voice of the Palestinian people. Its leader Yasser Arafat was invited to address the UN General Assembly, and international opinion, hitherto sympathetic to Israel on account of post-Holocaust guilt and admiration for having stood up to Arab military pressure, began to distance itself from the Jewish state.

Hizballah for its part, wanted to strengthen its political clout in Lebanon during the 1980s and become a power-broker in the country. To do this, it engaged in organisational branding by attacking Israel and the US, thereby earning the respect of Lebanese

Islamists, who were hostile to both countries. Since it could not afford to openly strike at either power using terrorist tactics (guerrilla operations being a different story), it resorted to carrying out attacks under a pseudonym: The Islamic Jihad Organization (IJO). The IJO was responsible for bombing American Marines in Beirut in 1983 (referred to in Chapter 2), and carrying out suicide bombings against Israeli interests in Argentina. One can see strong similarities here with LeT's frequent use of pseudonyms to internationally mask its involvement in cross-border attacks, even as it boasts about them domestically in Pakistan.

Both the PLO and Hizballah created an artificial distinction between their political and military wings, and further compartmentalised terrorist operations within a specific section of the military wing. In the PLO's case, major terrorist attacks during the 1970s were carried out by a network code-named 'Black September'. Consisting of PLO fighters who had ostensibly split from the mother organisation to engage in international terrorism, Black September was in fact the covert operations arm of the PLO, responsible for carrying out actions which the Palestinian leadership did not want to be directly associated with for fear of attracting Western condemnation. Hizballah similarly delegated terrorist operations to a sub-unit of its military wing, code-named the Special Research Apparatus (SRA).[5] Both Black September and the SRA were headed by charismatic terrorist leaders: Ali Hassan Salameh and Imad Mughniyeh, respectively. Both men were eventually killed by Israeli intelligence.

From what we have learnt about the relationship between ISI and LeT, it appears as though Pakistan is trying to replicate Middle Eastern conflict dynamics in South Asia. The operational-strategic model being used is a composite of PLO-Hizballah methods, and

[5] Bergman, *The Secret War with Iran*, p. 240.

therefore pushes forward two agendas at once: An external agenda of the Pakistani state, and a domestic agenda of LeT.

At the external level, the Pakistani state seeks to project its conflict with India as stemming from one solitary issue: The status of J&K. In the process, it avoids any mention of cross-border terrorism, which is after all, its sole instrument for ensuring that international attention does not fade away from J&K. Meanwhile, at the domestic or sub-state level, LeT wants to build up its street power and economic assets within Pakistan. Thus, events like 26/11 serve a dual purpose: They allow Pakistan to refocus international concern on J&K and they allow LeT to expand its influence at the expense of other *jihadist* groups, most of which are less loyal to the ISI. For the sake of keeping up appearances, the attacks themselves are carried out under pseudonyms to avoid triggering action against LeT directly, thereby allowing the group to continue operating in public.

There are, however, two weaknesses in the Pakistani strategy. First, major terrorist incidents cannot be obviously linked to the Kashmir issue, as that would strengthen India's argument that the Pakistani state is resorting to outright blackmail by attacking civilians. Unlike the Israeli-Palestinian conflict, the India-Pakistan stand-off is between two governments. Pakistan has less liberty to openly engage in underhand tactics that the PLO, as a stateless group, could get away with. Second, LeT cannot credibly portray itself as a Kashmiri organisation, because too much is already known about its overwhelmingly West Punjabi ethnic composition. There is no Kashmiri equivalent of Black September to claim that anti-Indian terrorism represents a popular movement among ordinary Kashmiris. These two factors mean that as long as India calibrates the nature of its response to major terrorist strikes, international attention will not automatically get drawn to J&K, but would instead stay focused on the more immediate problem of preventing such strikes.

Indian special operations need to reinforce, not contradict, this dynamic, which is presently working in our favour.

Learning from Israel's Mistakes

Unlike Israel, India can afford to wage a War of Attrition with Pakistan, if that is what Islamabad really wants. We have manpower resources to absorb the human costs that prolonged hostilities would entail. Our economic development would not get derailed by anything less than a nuclear strike, while attacks below this threshold can be adjusted to over a period of time. We can, therefore, respond to cross-border skirmishes without fearing the prospect of escalation. Even so, in our eagerness to punish Pakistan for its sponsorship of terrorism, we must avoid two mistakes that Israel made:

- *Wiping out a terrorist network, but further highlighting the political strength of its cause to foreign governments*: After the Munich Olympics massacre, Israel launched Operation Bayonet, an assassination campaign directed at Black September. The campaign was concentrated mainly in Western Europe, and had the tacit approval of host governments, who understood the Israeli imperative to exact revenge. However, it came with some costs. Firstly, the Israelis bungled an assassination in Norway. Believing that they were killing Ali Hassan Salameh, chief of operations for Black September, Mossad operatives gunned down an innocent Moroccan waiter. The embarrassing publicity that followed made clear that Israel was not merely a passive victim of terrorism, but was capable of using punishment methods not dissimilar from the terrorists'. Secondly, the assassinations mostly targeted second-rung Black September operatives, who were not directly connected with the Munich Massacre.[6] Not only did this raise questions about the

[6] Jonas, George, *Vengeance* (London: Harper Perennial, 2006), p. 113.

legitimacy of Israel's targeting process, but it showed just how extensive the network's presence was in Europe. European governments became wary about having to confront further acts of terrorism. Rather than get drawn into the Arab-Israeli conflict, they preferred to treat Black September as an Israeli problem that had landed on their doorsteps by accident.[7] (To protect American nationals in the Middle East, even the CIA backstabbed Mossad by entering into a no-aggression pact with Salameh. The Israelis were furious, but had no option other than to wait until the Black September chief was away from the protective umbrella of US intelligence, before killing him.)[8] Meanwhile, pressure on Israel to compromise with the Palestinians kept increasing. Thirdly, by crippling the PLO's terrorist arm, Israel deprived itself of an alibi to continue using 'hard' counterterrorism measures against the Palestinians. When the First Intifada broke out in 1987, the Israeli security community was wrong-footed because it could not portray the uprising as being part of a terrorist-orchestrated upheaval.

- *Relying on area penetration operations to the detriment of area dominance*: The second pitfall of Israeli counterterrorism

---

[7]A contrary interpretation of Operation Bayonet is that Israel had intended all along to send a message to European governments, and that Black September was only a convenient medium for this purpose. By killing terrorists on European soil, Mossad signalled to its foreign counterparts that if they did not take action against Palestinian networks, then it would act unilaterally. This would imply that the Israelis *wanted* the Europeans to know how widespread the Black September network was. In either case, the conclusion remains the same: Europe eventually did become more serious about combating terrorism, but may have still harboured silent grudges against Israel for compelling it to do so. Goldberg, Ari Ben, 'If you don't take care of this problem, we're going to have to', *Times of Israel*, 16 August 2012, http://www.timesofisrael.com/if-you-dont-care-of-this-problem-were-going-to-have-to-mossad-melman-raviv-spy-israel-iran/, 28 October 2013.

[8]Frohn, Axel, 'Hilfe vom Roten Prinzen', *Der Spiegel*, 8 October 2001, http://www.spiegel.de/spiegel/print/d-20289349.html, 29 October 2013.

relates to Hizballah. Between 1993 and 1999, Israeli troops and Hizballah militants engaged in constant cross-border skirmishing. To assist in intelligence collection and provide logistical support for SOF raids, Israel created a buffer zone in southern Lebanon, policed by members of a local anti-Hizballah militia. However, the militia turned out to be more of a liability than an asset, owing to the ease with which Hizballah counterintelligence penetrated it and 'turned' its members. In 1998, a double agent tricked a team of Shayetet 13 commandos into an ambush in which several were killed. After reviewing its losses, and recognising that its intelligence advantage over Hizballah was much smaller than its firepower advantage, Israel changed its policy in 1999.[9] Instead of launching area penetration missions, which were time-consuming to plan and rehearse for, and which took heavy casualties from positional defences in Hizballah strongholds, the Israelis switched to area dominance. They strengthened their incident response system and focused intelligence collection on detecting preparations for Hizballah attacks along the border. The pace of operational activity increased, but the granularity of intelligence improved since it was concentrated on monitoring militant movements a short distance inside Lebanon, which could be done using technical means from the Israeli side of the border. Hizballah had no solution to this, and its combat successes dwindled correspondingly. The shift from area penetration to area dominance at first appeared to be a defensive strategy, but it was actually just the opposite. Under area penetration, the responsibility for attacking Hizballah militants had rested with just a few elite units, while the rest of the Israeli army

[9] Byman, Daniel, 'The Lebanese Hizballah and Israeli Counterterrorism', *Studies in Conflict & Terrorism*, Vol. 34, No. 12 (2011), pp. 923-925.

remained in static defensive positions.[10] By switching to a more aggressive stance in territory controlled by Israel, the front line units actually increased the proactive component of Israeli policy towards Hizballah.

Israeli SOF are famous for their ability to conduct seemingly impossible missions, feats that are often credited to their offensive concept of operations. Lacking strategic depth, Israel prefers to fight wars on its enemies' territory rather than its own. However, frequent use of SOF against irregular adversaries has the disadvantage of prompting them to anticipate the damage that can be caused. The PLO initially used Black September to carry out acts of terrorism on a deniable basis, and then quietly shut down the group when its utility was over. Rather than Operations Bayonet and Spring of Youth (an Israeli commando raid in Beirut that killed three top terrorist leaders), it was decisions taken by the Palestinian leadership to make a tactical switch to dialogue that really weakened Black September.[11]

For its part, Hizballah has proven to be an exceptionally difficult target, since its ethnic composition differs greatly from that of the Palestinians. Israel has created special undercover military units, similar to Britain's 14 Intelligence Company and Force Research Unit, to gather intelligence in Palestinian areas. No comparable capability exists in the case of Lebanon. Although Hizballah has problems of its own, it has done a much better job of denying the Israelis an opportunity to target its leadership. Its organisational structure is also more fluid, thus ensuring that even the loss of an important commander can be weathered by rapid field promotions.[12]

---

[10] Byman, Daniel, *A High Price: The Triumphs & Failures of Israeli Counterterrorism* (New York: Oxford University Press, 2011), p. 239.

[11] Hoffman, Bruce, 'All You Need is Love', *The Atlantic Monthly* (December 2001), http://www.theatlantic.com/past/docs/issues/2001/12/hoffman.htm, 29 October 2013.

[12] Petrelli, Niccolo, 'The missing dimension: IDF special operations forces and strategy in the Second Lebanon War', *Small Wars & Insurgencies*, Vol. 23, No. 1 (2012), p. 67.

Special Operations Should Not Contradict Informal Cooperation Between Intelligence Services

By acting unilaterally in defence of its citizens, the Israeli state is well within its rights. However, it risks jeopardising intelligence cooperation with Arab governments whose support has hitherto helped suppress cross-border terrorism. A prime example is Jordan, which was Israel's closest partner in the Middle East until 1997, when Mossad attempted to assassinate a Palestinian activist living in the country.[13] Israeli-Jordanian relations never recovered from what the Jordanians perceived as a breach of trust. Given how dependent Israeli intelligence is on liaison channels with foreign partners, there is an inescapable need to factor in the damage that can be caused by carrying out assassinations on another state's sovereign territory. This applies to covert operations, which can always be disclaimed if they are exposed, and even more so to special operations, which are inherently attributable.

Even the 1976 Israeli raid on Entebbe, often praised as a classic commando raid, was crucially enabled by a liaison arrangement with Kenyan intelligence. Exploiting a family connection with one of the local Ugandan soldiers, a Kenyan officer infiltrated the building where Israeli hostages were being held, and gained valuable information about the terrorists' weapons and alertness level. His report contributed greatly to the mission intelligence package, and allowed the raiders to attain tactical surprise.[14] Another example of Mossad's dependence on intelligence liaison relates to Osama bin Laden. During the early 1990s, Mossad partnered with the intelligence service of a Gulf country to recruit bin Laden's personal secretary. The woman provided information which was used to plan an assassination attempt on the Saudi terrorist leader.

---

[13] Bar-Zohar, Michael and Mishal, Nissim, *Mossad: The Greatest Missions of the Israeli Secret Service* (New York: HarperCollins, 2012), pp. 266-278.

[14] Thomas, *Gideon's Spies*, p. 150.

Before the operation could be launched, however, relations between Israel and the partnering Arab state underwent a significant cooling off. Mossad could not proceed on its own with the assassination because it had not been able to obtain direct access to the secretary—all contact with her had been through the partnering service. By the time intelligence cooperation resumed, the secretary had moved out of bin Laden's inner circle.[15]

It is an open secret that post-26/11, India has benefited from intelligence sharing by friendly Arab states. The arrests of Abu Jundal, Fasih Mohammad and Abdul Karim 'Tunda' among others, were enabled by cooperation with countries where LeT has an active support network. Jeopardising this for the sake of retaliating to Pakistani delinquency along the LoC would be unwise. Instead, retaliation must be saved up for those rare occasions when Pakistan clearly violates the implicit limits that have been set upon its roguish behaviour. Another attack on the scale of 26/11, conducted by Pakistani nationals and traceable to the ISI-LeT combine, would be one such limit. Should such an event come to pass, India would have no alternative but to hit back. The only question is: How should it do so?

---

[15] Shpiro, Shlomo, 'Israeli Intelligence and Al Qaeda', *International Journal of Intelligence and CounterIntelligence*, Vol. 25, No. 2 (2012), pp. 242-243.

8

Synergising PSYOPS with Special Operations

This final chapter makes a straightforward argument: India should do the opposite of what Israel has done to combat terrorism. Israeli national security policy is tactically offensive but strategically defensive. The Jewish state fights for its very right to exist, in a region where terrorist groups (and once upon a time, even Arab governments) hoped to wipe it off the map. Its use of special operations is based on a national mentality that views mere survival as an accomplishment in itself.

India is in a very different position. Our geographic and demographic size, coupled with the strength of our armed forces, guarantee that we will successfully face down any challenge, overt or covert, from neighbouring states. It is not our political survival that is directly threatened by Pakistan-sponsored terrorism, but our economic prosperity that is indirectly threatened. We, therefore, must simultaneously conserve and harness our politico-military strength by being tactically defensive *and* strategically offensive. Rather than worry about everyday shooting incidents on the border, we need to concentrate on proactively shaping domestic, regional and global perceptions of Pakistan, in a manner that favours making no compromises with a terrorist state. Special operations are an important, indeed vital, component of this strategy. The strategy itself, however, would have to be primarily played out at the levels of diplomacy, media management and public communication.

For special operations to make a long-term contribution to Indian national security, they need to be synergised with psychological operations. This is the basic lesson which SOF mission planners can learn from their irregular adversaries. Terrorist and insurgent strategists know that, in the final analysis, they cannot hope to beat an established military power by force alone. So their strategy is oriented towards exhausting its will to fight, through non-kinetic means such as discourse-shaping.

Learning From the Tet Offensive

As an example, one need only look at what might have been the most strategically-decisive small unit operation of the Vietnam War: The storming of the US embassy in Saigon. Readers might need a little background to this event: By January 1968, US military spokesmen were proclaiming that the War was slowly being won, and that the years of building up South Vietnamese forces to fight communist insurgents were finally delivering results. Then, on 30 January 1968, the insurgents and North Vietnamese Army launched a coordinated series of urban assaults across the country. Having spent months inserting sleeper cells and building up arms caches in the middle of US-held territory, together they abruptly seized control of local strong points. After initial confusion and paralysis, the US and South Vietnamese forces counter-attacked and swiftly regained all the areas occupied.

The American public was not interested in the overall situation however: Its attention was fixated on just one location. The US embassy in Saigon, the capital of South Vietnam, had been attacked by a 19-man assault squad, in what would today be considered a *fidayeen* raid. Despite the fact that American servicemen elsewhere in Vietnam were doing a good job of fighting back, what television viewers in the US homeland saw was a national embarrassment unfolding at the embassy. Soldiers crouched behind walls and trees as the insurgents rampaged through the compound. Since the attackers could not enter the building (which had been locked down from the inside when the assault began), they remained in the open, firing at the security cordon. After a six hour firefight, they were neutralised.

More than any other combat action of the Tet Offensive (as the countrywide insurgent/North Vietnamese attack became known) the battle at the US embassy discredited all that had been said before about the war going well for the US and its allies. It caused bewilderment back on the home front. Ordinary Americans could not understand why if victory was indeed close, then how could the enemy get within striking range of the embassy? Contradictory and inaccurate media reporting magnified the sense of disillusionment, causing public commentators to openly question if the war was winnable. As soon as this question began to be asked, it assumed a life and momentum of its own: One that only ended when the US withdrew from the country in ignominy seven years later.

The lesson that can be drawn from this incident is: Special operations assume disproportionate significance when they discredit a dominant political narrative amongst the enemy. If we in India wish to use special operations for strategic effect against cross-border terrorism, then we need to closely study the narrative that the Pakistan Army, ISI and *jihadist* community are currently weaving together. From such analysis, we might be able to identify physical weaknesses whose exploitation would discredit the army/ISI/*jihadists* among their own supporters and reduce their political space.

Rather than satisfying our urge for vengeance, this is what a special operation should be aimed at: Reversing the enemy's strength and converting it into a hindrance which constrains his freedom to make further escalatory decisions. In the present context, this may require biding our time and constantly strengthening our area dominance capacity. Simultaneously we would need to build from scratch, a capability for area penetration. Such capability ought to be used only when the politico-ideological setting is conducive for generating a multiple shock effect. Instead of copying Israel or the US, we should find our own way, which will be better suited for our geostrategic situation.

Making Defence the Stronger Form of Warfare

For all their tactical aggressiveness, Israeli and American SOF have not succeeded in reducing the sense of insecurity felt by their respective countries. Strategically, both states have retreated behind security shields, viewing the outside through a prism warped by siege mentality. India does not need to be equally fearful, considering its successful record of integrating the world's largest Muslim minority, and repeatedly frustrating separatist movements backed by foreign powers. It can and should take a more assertive stance in South Asia, realigning regional power equations in a manner that leaves Pakistan both humiliated and isolated, as punishment for its support to terrorism.

Towards this, we should take a lesson from the Prussian Field Marshal Helmuth von Moltke. A leading military thinker of the nineteenth century, Moltke observed that although offense can be the stronger form of warfare in certain political and technological contexts, the preponderance of firepower and the lack of manoeuvrability on the modern battlefield made it suicidal.

Better to wait for the enemy to attack first, bleed him white, and only once he is reeling from heavy losses, close in rapidly and finish him off.[1] In terms of special operations, this means that India should concentrate on area dominance missions, where it has natural advantages in intelligence and logistics over the enemy. Only once these have been exploited to the fullest, should New Delhi devote time and resources to area penetration missions. The latter must be carried out when the enemy has already brought everything he has to the fight, and yet, has derived no real benefit. His morale and international credibility will be weakest at this point and like a brittle piece of wood, can be broken with one sharp blow.

Area Dominance: Not Glamourous But Essential

India's biggest failing in counterterrorism has been an inability to shift away from dependence on the Army, towards a police-centric model of combat operations.[2] During the worst years of terrorism in Punjab and J&K, 'islands of excellence' were developed in the local police forces, to confront and defeat terrorists on the latter's own terms. In Punjab, it was the Police Commandos while in J&K, it was the Special Operations Group. For the most part, however, the past successes of our police counterterrorist operations have been derived from good intelligence, not slick combat capability.

26/11 exposed this weakness. One commentary pointed out that the Mumbai policemen who faced the initial onslaught were not so much overcome by superior firepower, as they were undermined by their own poor marksmanship.[3] With no ammunition having been

---

[1] Krause, Michael D, 'Moltke and the Origins of the Operational Level of War', http://www.history.army.mil/books/OpArt/germany1.htm, 29 October 2013.

[2] Subramaniam, Arjun, 'Challenges of Protecting India from Terrorism', *Terrorism and Political Violence*, Vol. 24, No. 3 (2012), p. 405.

[3] LaRaia, William and Walker, Michael C, 'The Siege in Mumbai: A Conventional Terrorist Attack Aided by Modern Technology', in Haberfeld, MR and Hassell, Agostino von eds., *A New Understanding of Terrorism: Case Studies, Trajectories and Lessons Learned* (New York: Springer, 2009), p. 331.

provided for firearms training, this was hardly a surprise. Although the city police had an Anti-Terrorist Squad (ATS) of 450 men armed with sixty assault rifles, none were physically at hand when the attacks began. Nor could they be expected to be; the ATS is meant for offensive action against an opponent who wants to flee (such as mafia gangsters), and not those who stand and fight (like suicidal terrorists). They are not a highly-trained police Special Weapons and Tactics (SWAT) unit which can enter into a gunbattle from a position of initial disadvantage and then turn the situation around through superior fighting skill and technology.

In the years since, dedicated SWAT units have mushroomed across the country. We can take mild comfort from the fact that a mini-scramble for weapons and specialist assault equipment is underway. Even if most of this acquisition process is wasteful and geared primarily towards enhancing organisational budgets rather than addressing structural flaws, it also incrementally makes India a harder place for LeT *fidayeen* to operate in. Meanwhile, the Army is training its infantry Ghatak (commando) platoons to undertake counterterrorist assaults if asked by civilian authorities. The National Security Guard (NSG), which went into Mumbai bereft of basic equipment like night-vision devices, hands-free radios, corner-shot rifles and see-through-wall systems, has been massively expanded to include four regional hubs and is flush with new gear. Its authority to commandeer civilian aircraft in emergencies has been reiterated, thereby ensuring that deployment time would be cut down.

No Guarantees Can be Provided

That said, none of these measures can guarantee that the next major terrorist attack from Pakistan will not be as destructive as 26/11. The interrogation of LeT operative Daood Gilani alias David Headley has established that *fidayeen* attacks are professionally rehearsed, based on tactical data gathered by close reconnaissance of chosen targets. When a terrorist attack plan is being drafted, it takes into consideration the police capacity for immediate response, and the likely nature of that response. From 26/11, it would have become clear to LeT planners that remaining on the move is crucial to the combat effectiveness of *fidayeen* teams. As long as the various 'buddy pairs' in Mumbai stayed mobile, whether on the streets or in the two hotel complexes that were attacked, our security forces were unable to contain the damage they caused. Once the terrorists hunkered down into a positional defence, they lost the initiative and were slowly worn down by relentless pressure from the NSG. It is a safe bet that the next *fidayeen* strike would be similarly fluid in its targeting pattern: A mixture of fixed sites and targets of opportunity would be hit, before the gunmen settle into a defensive posture.

In any case, plugging the weakness in our urban SWAT capability is not the Army's job. While there is little doubt that our soldiers will react selflessly and promptly when called out on a counterterrorist mission in peacetime locations, it is actually the local police who should take the lead in handling the situation. In legal terms, the police would be better protected if they use lethal force in situations where it is not immediately clear who the enemy is—foreign terrorists or Indian citizens brainwashed to commit acts of terrorism. However, the Army can still help by conducting joint drills with the police to forge cross-agency contacts, which will certainly be needed in a crisis.

The real challenge in preventing another 26/11 does not lie in the SWAT realm. Rather, it is a problem of surveillance. LeT has found a way to outflank our well-guarded land borders through using the sea route for infiltration. What has proven successful once will certainly be attempted again. Besides bringing economic targets and defence research and development (R&D) facilities in southern India within LeT's reach, the maritime dimension also makes coastal areas with a strong presence of foreigners highly vulnerable. After the international backlash that followed the 2008 attack, LeT is unlikely to directly target foreign tourists in India. Instead it is urging Indian Mujahideen members to do so. Stopping a LeT/IM firearms assault on a civilian location, such as a beach resort, would require creating pockets of intensive human and technical surveillance around potential target clusters.

To sequentialise the creation of these pockets, in order of highest threat priority, intelligence analysts could examine jurisdictional boundaries—LeT will certainly plan to conduct its next major *fidayeen* raid along seams in our surveillance and incident response system. For this reason, an attack on shipping in the high seas is also highly likely, since it would not be clear in the initial moments, which agency should take the lead in responding. Regardless of what specific form the next attack takes, our police/army/navy response must be prompt and fierce. During 26/11, the US and UK used the Mumbai police's poor reaction to the attacks, as an excuse to divert international attention from Pakistan's role. We cannot give these countries another chance to airbrush Pakistani sponsorship of cross-border terrorism, by appearing to be ham-fisted in our own domestic area dominance operations.[4]

Hitting Back

Once an attack has occurred, despite all reasonable precautions and countermeasures on our part, the need to retaliate will be inescapable. We have learnt through bitter experience that some

[4] Interestingly, when Israel was hit by a seaborne Palestinian *fidayeen* raid on Tel Aviv on 6 March 1975, its security response was as poor as ours on 26/11. During both attacks, local police initially thought that a gang war had broken out. They could not believe that what was unfolding was a terrorist strike. Eleven people were killed in the Tel Aviv attack, including three soldiers trying to free hostages who had been seized at the Savoy Hotel. Like Mumbai 2008, the 1975 attack took place on the eve of peace talks between Israel and Egypt, leading some commentators to hurriedly conclude that it was aimed at sabotaging the talks, and hence Israel should exercise restraint. [Anonymous], 'The Savoy Attack', *Journal of Palestine Studies*, Vol. 4, No. 4 (1975), pp. 127-129.

Western governments in particular are playing a double game: Vaguely admonishing Pakistan for not acting against terrorism, while refusing to admit that Pakistani state agencies are complicit in terrorist attacks. As Major General (Retd) GD Bakshi has noted, India is being sacrificed in exchange for the ISI's cooperation in busting *jihadist* conspiracies directed at Western homelands.[5] Racism thus remains at the root of Anglo-American policy on cross-border terrorism directed against India. No amount of political bonding over shared democratic values or growing business opportunities can override this dynamic.

Getting the Legal Framework Right

So here comes the big Question: Faced with an indifferent international audience that ignores its suffering and righteous anger against Pakistan, what can India do? A possible answer: It can cease waiting for the US and UK to develop a conscience and instead take unilateral punitive action, after carefully laying the political and diplomatic groundwork to justify conducting reprisals. According to a legal precedent laid down in 1928, a cross-border reprisal is legitimate if it is:

- precipitated by a prior act of illegal aggression,
- preceded by an unmet demand for peaceful restitution,
- proportionate to the initial act of aggression and,
- discriminate in ensuring that innocent third parties are not injured.

The problem is that a very small window of opportunity exists to link an act of reprisal to a previous act of terrorism. One expert estimates that the window lasts no more than seventy-two hours. After that, both public and international opinion begin to disassociate one act of aggression from the other, such that the injured party

[5]Bakshi, GD, 'Mumbai Redux: Debating India's Strategic Response Options', *Journal of Defence Studies,* Vol. 3, No. 4 (2009), p. 22.

can actually end up looking like the perpetrator. Also, to generate a cathartic effect upon the populace of the reprising state, the act of reprisal must be publicly acknowledged, which means that it cannot be covert.[6] Therein lies the real policy dilemma facing Indian strategists: From a military perspective, it makes sense to retaliate covertly after a lengthy gap, when the enemy's vigilance has been lowered, at a time and place of our choosing. From a political and diplomatic perspective however, the act of retribution should be overt and occur as soon as possible after the terrorist attack, in order to show a cause-effect relationship.[7]

Lengthening the Timeframe of Retaliation

India, thus, needs to strike a delicate balance between operational and psychological considerations, when planning an area penetration mission into Pakistan. As a first step, it needs to use international laws and treaties on counterterrorism to highlight Islamabad's continuing perfidy and reluctance to follow through on past commitments. In this regard, the 2004 bilateral agreement that no Pakistani territory would be used to shelter terrorists sits awkwardly with LeT's proven involvement in 26/11 and Pakistan's ongoing protection of this group. Holding up Pakistan's failure to deliver on its past promises would help us establish a pattern of trust betrayal. It would also expand the political timeframe for conducting legitimate reprisals in the aftermath of future attacks. Instead of three days, our aim should

---

[6] Livingstone, Neil C, 'Proactive Responses to terrorism: Reprisals, Preemption and Retribution', in Livingstone, Neil C and Arnold, Terell E, *Fighting Back* (Lexington, MA: Lexington Books, 1986), pp. 122-123.

[7] Generally, international opinion does not harshly condemn an act of retaliation, if it is carried out soon after a grave provocation, and against a military target. Under no circumstances however, does it come out in favour of such retaliation, except in highly unusual cases like 9/11 and that too, only because the US was the aggrieved party. See Falk, Richard A, 'The Beirut Raid and the International Law of Retaliation', *American Journal of International Law*, Vol. 63, No. 3 (1969), p. 63.

be to engineer an international mood that accepts a retaliation occurring even as late as three months after a terrorist attack as 'acceptable'. This perceptual shift in foreign government and public opinion can only be brought about if our diplomats, political leaders and academicians remain proactive in reminding neutral countries about the danger of Pakistani terrorism.

To be fair, we have not done too badly in this regard. Pakistani commentators had believed during 2009-11 that international pressure would force India to 'forget' the terrorist carnage in Mumbai and restart peace talks that would focus on J&K. In the long run, their complacency might have even been justified, had Osama bin Laden not been found in May 2011 hiding on the edge of a major Pakistani military base. Literally overnight, global views of Pakistan became a lot more sceptical, thereby vindicating our persistent argument that we were faced with a state that harbours terrorist groups.

Pakistan has since found it harder to bury memories of its past misdeeds than it had anticipated. Even five years after the 26/11 attack, US President Barack Obama made a point to ask Prime Minister Nawaz Sharif, why the trial of LeT suspects had not yet begun. In 2012, the US put a USD 10 mn bounty on LeT chief Hafeez Saeed, calling for evidence that would help convict him of international terrorism. Although these measures were cosmetic, and it would be a mistake to read too much meaning into them, one thing is clear: Washington has been signalling to Islamabad that terrorism is not an acceptable tool of statecraft in the post-Cold War era. This is important, because it denies Pakistan the means to convert its covert paramilitary operations into overt diplomatic payoffs, as was done with the Soviets in Afghanistan. Even though the US has adopted a position of careful (and immoral) neutrality on the issue of cross-border terrorism as concerns India, such a stance still helps to preserve the regional status quo, which is not what Pakistan wants. After all, what is the point of sponsoring terrorism if you do not derive some strategic benefit from it?

Selecting the Target(s)

Before saying what India should do in response to a terrorist attack, it is important to note what India should not do: It should not carry out a special heli-borne operation (SHBO) against *jihadist* training camps in Pakistan Occupied Kashmir (PoK). There are three reasons why such a move would be counterproductive.

Firstly, in the immediate aftermath of a big attack by LeT, the Pakistani army swiftly evacuates *jihadist* camps in PoK. It dresses up this move as some form of counterterrorist action, but the real purpose is clear—to deny us a fixed and legitimate target against which to strike. Secondly, hitting a site on PoK real estate would have the negative side effect of creating a linkage between terrorism and the Kashmir issue, which is exactly what Pakistan wants in the long run. Islamabad hopes that over time, foreign governments will start to believe that an equitable 'solution' to Kashmir (i.e., one where India being the larger but non-aggressive power is forced to make all the concessions) would be a guarantee for regional peace. We must not play into Pakistan's hands by allowing world attention to be split between terrorism and J&K. Instead, the focus of international debate must remain on tracking and curtailing *jihadist* activities in Pakistan proper. Thirdly, it is virtually assured that even if India were to carry out a SHBO, the Pakistani army would have anticipated the ingress route, the tactics we would use on the ground, and the exit plan. Even if they cannot stop us from entering their airspace, for lack of adequate reaction time, once the assault is underway they will have ample opportunity to launch a ground and aerial counter-attack. Since the SHBO would anyway take place at a time of heightened tensions following a major terrorist outrage in India, they would be extra vigilant to anomalies in our air traffic pattern over J&K. Thus, tactical surprise even if attained, would be too small to override the enemy's ability to interfere with our special operation. We would suffer heavy casualties. This is not to say that risk aversion should dominate our mission planning, but we must deny the enemy any chance to inflict further harm on us, beyond what he would already have done.

Instead of hitting training camps in PoK, why not hit LeT and Jamaat-ud Dawa (JuD) infrastructure in Pakistani Punjab? According to a 2013 study published by the US Army's Combating Terrorism Center, around 90 per cent of all LeT terrorists are Punjabis. The bulk of them come from just ten districts, of which three—Gujranwala, Faisalabad and Lahore, host many of the group's fixed assets.[8] Admittedly, the logistics of attacking these places are a problem, since Lahore has a massive Pakistan Army presence and the other two districts do not lie adjacent to the international boundary. Even so, substitute targets might be found in other, more vulnerable border districts such as Kasur, Bahawalnagar, Bahawalpur and Multan. If all else fails, there is Karachi to the far south. Pakistan can learn the hard way that what goes around comes around. Its naval Special Services Group (SSG) had helped to train the 26/11 attackers. In the future these same training facilities can come under seaborne attack, subsequently claimed by unknown elements within the Pakistani Taliban.

Targeting the JuD alongside LeT would make sense, since the two organisations share personnel, offices, bank accounts and telephone lines. In effect, they are one and the same, with an artificial

---

[8] Rassler, Don; Fair, C Christine; Ghosh, Anirban; Jamal, Arif and Shoeb, Nadia, *The Fighters of Lashkar-e-Taiba: Recruitment, Training, Deployment and Death*, Combating Terrorism Center, pp. 26-28, http://www.isn.ethz.ch/Digital-Library/Publications/Detail/?id=163497, 8 November 2013.

distinction being drawn primarily for international consumption.[9] Likewise, hitting the SSG along with the ISI would be justified, since both are directly involved in training LeT *fidayeen* for attacks on India. Whether our retaliation should just be confined to active LeT members, or should include their immediate families, is a more problematic question. It has been estimated that around 40 per cent of all LeT terrorists are recruited through family networks, thus making a mockery of any notion that the relatives of terrorists should be automatically classed as non-combatants and left alone.

On the other hand, there is a generally-observed norm in counterterrorism to avoid inflicting collateral damage except in rare cases, when the payoff of attacking regardless of civilian deaths is too great. Baitullah Mehsud, the leader of the Pakistani Taliban, was killed along with his wife and in-laws in 2009 by a US drone strike. During the 1980s and '90s, the CIA and Mossad each killed a brother of Hizballah terrorist Imad Mughniyeh, hoping to smoke him out of hiding. Mossad actually intended to lure Mughniyeh to attend his brother's funeral, where it planned to kill him and everyone else who was present, through a helicopter missile strike. So, even if killing a terrorist's close family members is something to be generally avoided, it is nonetheless accepted within the intelligence world that there are occasions when such measures will be necessary. We should consider this.

Play the Enemy by the Rules that He Has Chosen

India stands at the threshold of a moral dilemma. If it seeks to limit its retaliation only to those specific LeT cadres and installations that were actually involved in planning a terrorist strike, probably no retaliation shall occur. This is because even the best intelligence agency cannot hope to track down a man in real time, when he is

[9] Fair, C Christine, 'Lashkar-e-Tayiba and the Pakistani State', *Survival*, Vol. 53, No. 4 (2011), pp. 3-4.

hiding with a hostile country with support from local intelligence services. To appreciate this point, one needs to only look at three of Mossad's most famous manhunts: Ali Hassan Salameh (mastermind of the 1972 Munich Olympics massacre); Khalid Al-Wazir (planner of the 1975 Tel Aviv Savoy Hotel raid—a close prototype of 26/11); and Imad Mughniyeh (responsible for killing more Westerners than any terrorist prior to 9/11).

Mossad took six years to get to Salameh in Lebanon, fifteen to kill Al-Wazir in Tunisia and more than two decades to assassinate Mughniyeh in Syria. In Mughniyeh's case, it had the help of numerous Western intelligence agencies, but still had to wait until a sighting of him surfaced on its informant web. This is the same problem that we face with LeT operational planners: We know who they are, but not where they are in real time.

Besides operational logic, perhaps the most compelling case for our hitting a broad range of targets has been made by LeT itself, when it planned the Mumbai attack. According to David Headley, the decision to target Nariman House, the city's Jewish cultural centre, was part of a deliberate effort to link the India-Pakistan conflict with wider *jihadist* dynamics in the Middle East. So, if LeT wants to do business as per targeting norms applied in the Arab-Israeli conflict, it would be discourteous of us not to reciprocate. Hitherto, we have avoided hitting at soft targets in Pakistan (or even perfectly legitimate hard targets, such as military installations) for the sake of maintaining peaceful relations. However, if Pakistan, through its ISI, SSG and LeT, is committed to wrecking the peace with India, then it should be prepared to face the costs. Attacking soft targets connected with the Pakistani military or LeT would be a perfectly justified proposition under such circumstances.[10]

[10] It would be advisable to consider ISI 'forward sections' and SSG installations as reserve targets, that could be hit should either the Pakistani state or LeT choose to escalate by carrying out further terrorist attacks on India.

Needed: A capacity for Area Penetration

India is a long way from having the kind of special operations capability needed to conduct strikes inside Pakistani Punjab and Sindh during peacetime. Our entire worldview on cross-border terrorism has been built upwards from a 'war paradigm'. This paradigm was the foundation of force postures and doctrines that were first developed in the 1980s, when the ISI wreaked havoc through the Khalistani militancy. Beginning with notions about massive armoured strikes across the western desert that would severe Pakistan in half (the Sundarji Doctrine) and continuing post 2001 to the so-called Cold Start Doctrine, we have viewed war as a fitting answer to Pakistani sponsorship of terrorism.

Though there is little doubt that Pakistan is an uncivilised state that ought to be vivisected once again, we cannot escape the strategic reality that it possesses nuclear weapons and would not hesitate to use them.[11] If our decision-makers are presented with a stark choice between backing a special operation that could potentially escalate into nuclear war, or stomaching whatever terrorist outrages Pakistan dishes out, they will choose the latter option. History has already demonstrated that our instinct for self-preservation is stronger than our sense of anger, no matter how badly we are provoked. Operation Brasstacks (1987) was viewed by the ISI as a humiliating climb-

[11] As early as 1948, Bhola Nath Mullick, then an Assistant Director in the Intelligence Bureau, wrote an assessment for Prime Minister Jawaharlal Nehru pointing out that the Pakistani invasion of Jammu and Kashmir the year before was unlikely to be a one-off event. Until and unless the state of Pakistan was broken up, and its foundational ideology discredited, the country would remain an existential threat to India. The intensity of the threat diminished after the formation of Bangladesh in 1971, but continues to this day in a reduced form. Unfortunately, another war to break up Pakistan is no longer feasible, and India's only real option is to conduct covert operations towards this end, while hoping that internal divisions within Pakistan cross the breaking point. Banerjee, Sashanka S, *A Long Journey Together: India, Pakistan and Bangladesh* (Charleston, South Carolina: Booksurge, 2008), pp. 204-205.

down for India in the face of Pakistan's nuclear arsenal.[12] The same applies to Operation Parakram (2001-02). The fact that we did not make similarly threatening moves after 26/11 would doubtless have strengthened their belief that nuclear-weapon enabled terrorism is a viable tool of foreign policy.

Explore the SOF Option, Alongside Conventional Military and Nuclear Deterrence

Therefore, we must begin planning an area penetration mission with the understanding that if it fails, it *may or may not* be backstopped by a conventional military attack. There can be no automatic assumption that the enemy would just be overawed by our public anger and allow us to move unhindered into his airspace. A SHBO would certainly be intercepted by fighter aircraft, or shot down by air defence missile batteries. Intense American pressure might cause us at such a point to avoid sending in our own fighters to pound Pakistani military bases. If we were to suspend hostilities halfway, once Pakistan has fired on our troops, the net result would be a partially successful mission. India would indeed have struck back, but with heavy costs to our credibility as a military giant.

Besides investing in a target-specific and task-specific SOF capability, India also needs to maintain its conventional military advantage over Pakistan on land, sea and air. Pakistan has to be forced to understand that its nuclear weapon-enabled terrorism can be fought away from public view with our own nuclear weapon-enabled counterterrorism. If the Pakistani military thereafter chooses to escalate by going in for a conventional counter-retaliation, India would have the ability to match this.

To argue that even a rogue entity such as the Pakistan Army would respond to the deaths of LeT terrorists with nuclear weapons

[12] Bearden, Milt and Risen, James, *The Main Enemy: The Inside Story of the CIA's Final Showdown with the KGB* (New York: Ballantine Books, 2003), pp. 263-264.

strains credulity beyond breaking point. The LeT's middle-rank leadership may be a strategic asset of the Pakistani Deep State, but like all covert action instruments, it too is dispensable once it causes more trouble than it is worth. Furthermore, as the Pakistanis never tire of pointing out to Western interlocutors, India and Pakistan face a common threat in terrorism. Ergo, any counterterrorist action by India is consistent with Pakistan's own interests—our diplomatic machinery should repeatedly strive to impress this point upon both Western and Arab governments. While no foreign power can be expected to support India, at least their expressions of hostility can be substantially muted by sound politico-diplomatic preparation on our part.

Target Identification and Preparation

To enjoy any real chance of success, a cross-border attack by Indian special operations forces has to:

- be independent of air power (except for overall command and control and for 'hot' extraction under enemy fire),
- involve the use of disguise and subterfuge to infiltrate past front line defences,
- hit at targets that rely on anonymity rather than physical security for their protection.

In short, we cannot employ the Anglo-American model of special operations, whose effectiveness is heavily predicated on having air superiority. Close air support and tactical airlift are two luxuries that our SOF would have to do without, while operating against terrorist infrastructure in Pakistan. This requires that we resort to more clandestine methods of infiltration and attack, and go for semi-protected or unprotected targets that are nevertheless closely linked to LeT/ISI/SSG. Finding such targets is an intelligence responsibility that should rest with the Research & Analysis Wing (RAW) and the Directorate General of Military Intelligence. Scouting the targets' vulnerability to attack should however, be an SOF responsibility. We urgently need a unit like the US Army's ISA, to conduct deep-penetration close target reconnaissance in hostile territory.

The best organisation to house such a unit would be the Special Group of the Cabinet Secretariat's Establishment 22. Since this ultra-secret organisation already has experience of working with émigré fighters, it could provide a platform for covertly training Pakistani dissidents in commando tactics. In the event that a direct action mission needs to be carried out by Indian SOF against a high-value terrorist infrastructure site in Pakistan, these dissident fighters would be in a position to supply the ground-level data necessary for assault planning. Their involvement would probably not extend to logistical support for the actual mission (since operational security would be put at risk). However, general preparation for area penetration special operations, such as creating weapon caches and identifying potential lying-up posts in the target areas, could be entrusted to them. Just as Pakistan is using the so-called Indian Mujahideen to mask its proxy war against India from Western audiences, so too can India use Pakistani rebels to obfuscate its counter-proxy war against Pakistan. For this to happen however, a clear policy and bureaucratic consensus needs to first be forged that such a counter-proxy war is necessary for India's security. Nowhere does such a consensus appear to exist.

Focus on Intelligence and Logistics, by Better Tasking and Training

At the moment, India does not even have a military unit similar to Tier-1 American SOF, to handle direct action missions based on intelligence gathered by Central government agencies. There is no equivalent of Delta Force, Naval Special Warfare Development Group (DEVGRU—better known as SEAL Team 6), or the British SAS. Even if such units had existed in the Indian Army, they would be of little use for the kind of missions that need to be carried out. Instead of direct action, what we really need is first to build up our covert logistics and intelligence capability for operating deep inside Pakistan. Towards this, tasking and training of our SOF need to be revamped.[13]

Our current SOF capabilities are optimised for area dominance in a rural counter-insurgency role. We are still somewhat lacking in counterterrorist SWAT capacity (which is needed for urban area dominance) and completely lacking in area penetration capability. Whatever operations have been carried out by the Indian Army across the LoC, have been shallow-penetration raids and reconnaissance missions.[14] While there is no doubt that Army Special Forces (particularly 9 PARA SF and the Establishment 22 Special Group) have generated valuable information from visual observation, document seizures and prisoner snatches in PoK, these efforts need to be greatly enhanced by firm political mandates for aggressive operations. It would perhaps be best if the ultimate responsibility for operational oversight rested with the National Security Advisor (NSA). Being in direct contact with the Prime Minister and holding a rank equivalent to Minister of State, the NSA would be the best authority to evaluate the strategic ramifications of a proposed

[13] As Brigadier (Retd) Deepak Sinha observed at a CLAWS seminar (held on 17 December 2013), the Western special operations model is optimised for clandestine operations, not covert operations. Rather than being deniable, activities by Western SOF are only meant to remain secret for as long as they last, after which they might be owned up to. There is no political imperative for such operations to remain permanently untraceable. If India were to emulate the Western model of training and deployment, it would end up with a SOF capability built at great cost, but useless for the purpose of covertly striking at Pakistan on a continuing basis.

[14] One Indian Army officer has noted that the 'strategy for India is a plus minus twenty kilometre operation'. Long-range strikes, upto a depth of 200 kilometres or more inside Pakistan, simply do not feature in Indian counterterrorist planning. Quoted by Singh, Narendra (Brig), 'Interactive Session', in Oberoi, Vijay ed., *Special Forces: Doctrine, Structures and Employment Across the Spectrum of Conflict in the Indian Context* (New Delhi: Knowledge World, 2006), p. 137.

special operation.[15] His support, or the lack of it, would also serve as an indicator of the strength of our own national consensus on combating cross-border terrorism, and prevent strategic overreach.

Lastly, our selection and training model would not suffice for the tasks that need to be carried out. Instead of creating highly fit paratroopers (which is what some of our SF personnel effectively are) we need to deep-select officers and men with linguistic skills. Blending into a locality of urban Pakistan will be much tougher than trekking through the mountains of PoK to photograph terrorist camps. Subtle differences in dialect would be all that is required to compromise an SOF operator in hostile territory.[16] To enhance their chances of survival, men (and perhaps women) chosen for area penetration missions inside Pakistan need to be intensively trained in close and unarmed combat skills, as well as evasive driving. Their instructional curricula would be more akin to that used by VVIP close protection units, rather than a traditional military SOF unit. Instead of physical fitness, their main attribute would have to be the ability to remain inconspicuous while reconnoitring a target.

Make War Covertly, Talk Peace Overtly

In the final analysis, India needs to fight Pakistan on the terms chosen by Pakistan—those of special operations with a strong covert component. It must, however, not allow a disconnect to appear between its covert paramilitary activities and the declared policy of the Indian government. This has been Pakistan's mistake—it claims to oppose terrorism, but then does everything possible to support

---

[15] Lidder, HS, 'Employment of Special Forces to Achieve Strategic Objectives: Exploiting Opportunities with a Proactive Approach', in Yadav, VS ed., *Employment of Special Forces: Challenges and Opportunities for the Future* (New Delhi: Vij Publications, 2012), p. 94.

[16] Bailey, Bill, 'Hearts and Minds, Pseudo Gangs and Counter Insurgency: Based upon Experiences from Previous Campaigns in Kenya (1952-60), Malaya (1948-1960) & Rhodesia (1964-1979)', published in the *Proceedings of the 1st Australian counter Terrorism Conference, Edith Cowan University, Perth Western Australia, 30 November 2010*, p. 13.

it, and worse, regularly gets found out in the process. Having failed to win a single war with India through open hostilities, Pakistan is increasingly turning to self-contained acts of aggression carried out through proxies to project its will onto India. It rightly supposes that international pressure will force India to step back from launching overt military retaliation. What India now needs to do is find a way of playing Pakistan at its own game—making war covertly while talking peace overtly.

This book has found, through its analysis of five irregular wars (the Franco-Algerian War, the Vietnam War, the Arab-Israeli conflict, the Rhodesian Bush War, the Soviet-Afghan War, and the Northern Ireland conflict) that special operations can play a crucial role in shaping strategic outcomes. For this to happen, they need to be integrated into national strategy at the highest political level. Furthermore, they should be fully synchronised with the government's efforts to shape public and international opinion. This means that they should not be used for strategically defensive purposes, as favoured by the Anglo-American model, but rather for offensive purposes, as seen in the Germanic model. Special operations are a means of communicating with irregular adversaries and terrorist states in a language which these are already fluent in. The need for such communication is undeniable.

Bibliography

1. Andrade, Dale, 'The Vietnam War's Secret Side', *International Journal of Intelligence and CounterIntelligence*, Vol. 14, No. 4 (2001), pp. 582-586.
2. [Anonymous] 'After Munich: Israel Debates the Response', *Journal of Palestine Studies*, Vol. 2, No. 2 (1973), pp. 142-149.
3. [Anonymous], 'The Savoy Attack', *Journal of Palestine Studies*, Vol. 4, No. 4 (1975), pp. 127-132.
4. [Anonymous], 'The Soviet Elite: Spetnaz', in Lewis, Jon E ed., *SAS and Elite Forces* (Bristol: Parragon, 1996), p. 50.
5. Arreguin-Toft, Ivan, 'How the Weak Win Wars: A Theory of Asymmetric Conflict', *International Security*, Vol. 26, No. 1 (2001), pp. 93-128.
6. Baev, Pavel K, 'How Afghanistan Was Broken: The disaster of the Soviet intervention', *International Studies Review*, Vol. 15, No. 3 (2012), pp. 249-262.
7. Bailey, Bill, 'Hearts and Minds, Pseudo Gangs and Counter Insurgency: Based upon Experiences from Previous Campaigns in Kenya (1952-60), Malaya (1948-1960) & Rhodesia (1964-1979), *Proceedings of the 1st Australian counter Terrorism Conference*, https://www.ro.ecu.edu.au/cgi/viewcontent.cgi?article=1000&context=act, 30 November 2010.
8. Bakshi, GD, 'Mumbai Redux: Debating India's Strategic Response Options', *Journal of Defence Studies*, Vol. 3, No. 4 (2009), pp. 16-26.

9. Ball, Rhys, 'The strategic utility of New Zealand Special Forces', *Small Wars & Insurgencies*, Vol. 22, No. 1 (2011), pp. 119-141.
10. Balor, Paul, *Manual of the Mercenary Soldier: A Guide to Mercenary War, Money, and Adventure* (Boulder, CO: Paladin Press, 1988).
11. Banerjee, Sashanka S, *A Long Journey Together: India, Pakistan and Bangladesh* (Charleston, South Carolina: Booksurge, 2008).
12. Bar-Zohar, Michael and Mishal, Nissim, *Mossad: The Greatest Missions of the Israeli Secret Service* (New York: HarperCollins, 2012).
13. Bearden, Milt and Risen, James, *The Main Enemy: The Inside Story of the CIA's Final Showdown with the KGB* (New York: Ballantine Books, 2003).
14. Beckwith, Charlie A, *Delta Force: The Army's Counterterrorist Elite Unit* (New York: Avon, 1983).
15. Bergen, Peter and Cruickshank, Paul, 'Revisiting the Early Al Qaeda: An Updated Account of its Formative Years', *Studies in Conflict & Terrorism*, Vol. 35, No. 1 (2012), pp. 1-36.
16. Bergman, Ronen, *The Secret War with Iran* (London: Oneworld, 2008).
17. Biddle, Stephen D, 'Allies, Airpower, and Modern Warfare: The Afghan Model in Afghanistan and Iraq', *International Security*, Vol. 30, No. 3 (2005), pp. 161-176.
18. Bowden, Mark, *Black Hawk Down* (London: Corgi, 2000).
19. Brailey, Malcolm, *The Transformation of Special Operations Forces in Contemporary Conflict: Strategy, Missions, Organisation and Tactics* (Duntroon, Australia: Land Warfare Studies Centre, 2005).
20. Braithwaite, Rodric, *Afgantsy: The Russians in Afghanistan 1979-89* (London: Profile, 2012).
21. Burli, RG, 'Offensive Air Power in the high Mountains', *Air Power*, Vol. 2, No. 1 (2007), pp. 87-105.
22. Byman, Daniel, 'The Lebanese Hizballah and Israeli Counterterrorism', *Studies in Conflict & Terrorism*, Vol. 34, No. 12 (2011), pp. 917-941.
23. Byman, Daniel, *A High Price: The Triumphs & Failures of Israeli Counterterrorism* (New York: Oxford University Press, 2011).
24. Carney, John T and Schemmer, Benjamin F, *No Room for Error: The Covert Operations of America's Special Tactics Units from Iran to Afghanistan* (New York: Random House, 2003).

25. Cassidy, Robert M, 'Counterinsurgency and Military Culture: State Regulars versus Non-State Irregulars', *Baltic Security and Defence Review*, Vol. 10 (2008), pp. 53-85.
26. Cassidy, Robert M, *Counterinsurgency and the Global War on Terror: Military Culture and Irregular War* (Stanford, CA: Stanford Security Studies, 2008).
27. Charters, David A, 'Special Operations in Counter-Insurgency: The Farran case, Palestine 1947', *RUSI Journal*, Vol. 124, No. 2 (1979), pp. 56-61.
28. Chatterjee, Debalina, 'Rashtriya Rifles in Kashmir: India's Counter-terrorist Force', *Air Power*, Vol. 7, No. 1 (2012), pp. 95-116.
29. Church, George J, 'The Secret Army', *Time*, 31 August 1987, pp. 12-14.
30. Cilliers, JK, *Counter-Insurgency in Rhodesia* (London: Croom Helm, 1985).
31. Cleave, Michelle Van, 'Strategic Counterintelligence: What Is It and What Should We Do About It?', *Studies in Intelligence*, Vol. 51, No. 2 (2006), accessed online at https://www.cia.gov/library/center-for-the-study-of-intelligence/csi-publications/csi-studies/studies/vol51no2/strategic-counterintelligence.html, on 8 November 2013.
32. Clegg, Will, 'Irregular Forces in Counterinsurgency Warfare', *Security Challenges*, Vol. 5, No. 3 (2009), pp. 1-25.
33. Cline, Lawrence E, *Intelligence, Special Operations and Counterterrorism: Improving the Links*, Paper delivered at the International Studies Association Annual Convention, March 2004.
34. Cline, Lawrence S, 'Special Operations and the Intelligence System', *International Journal of Intelligence and CounterIntelligence*, Vol. 18, No. 4 (2005), pp. 575-592.
35. Cohen, Hillel and Dudai, Ron, 'Human Rights Dilemmas in Using Informers to Combat Terrorism: The Israeli-Palestinian Case', *Terrorism and Political Violence*, Vol. 17, Nos.1-2 (2005), pp. 229-243.
36. Coll, Steve, *Ghost Wars: The Secret History of the CIA, Afghanistan and Bin Laden, from the Soviet Invasion to September 10, 2001* (New York: Penguin, 2004).

37. Cooley, John K, *Unholy Wars: Afghanistan, America and International Terrorism* (New Delhi: Penguin, 2001).
38. Connelly, Mark and Willcox, David R, 'Are You Tough Enough? The image of the special forces in British popular culture, 1939-2004', *Historical Journal of Film, Radio and Television*, Vol. 25, No 1 (2005), pp. 1-25.
39. Constanza, William, 'Hizballah and Its Mission in Latin America', *Studies in Conflict & Terrorism*, Vol. 35, No. 3 (2012), pp. 193-210.
40. Couch, Dick, *The Warrior Elite: The Forging of SEAL Class 228* (New York: Three Rivers Press, 2001).
41. Crawford, George A, *Manhunting: Counter-Network Organization for Irregular Warfare* (Hurlburt Field, Florida: Joint Special Operations University, 2009).
42. Crist, David, *The Twilight War: The Secret History of America's Thirty-Year Conflict with Iran* (New York: Penguin, 2012).
43. Darwin, John, *After Tamerlane: The Rise and Fall of Global Empires, 1400-2000* (New Delhi: Penguin, 2008).
44. Datta, Saikat, 'Air Power and Special Operations', *Air Power*, Vol. 2, No 2 (2005), pp. 91-117.
45. Davies, Nicolas JS, 'Evidence of an American Dirty War in Iraq', *Peace Review: A Journal of Social Justice*, Vol. 19, No. 3 (2007), pp. 435-443.
46. Dobbie, Elliot VK, 'The Word "Commando"', *American Speech*, Vol. 19, No. 2 (1944), pp. 81-90.
47. Doohovskoy, Andrei A, *Soviet Afghan War Revisited: Analyzing the Effective Aspects of the Counterinsurgency Effort*, MA Thesis presented at Harvard University in September 2009.
48. Downes, Alexander B, 'Draining the Sea by Filling the Graves: Investigating the Effectiveness of Indiscriminate Violence as a Counterinsurgency Strategy', *Civil Wars*, Vol. 9, No. 4 (2007), pp. 420-444.
49. Fair, C Christine, 'Lashkar-e-Tayiba and the Pakistani State', *Survival*, Vol. 53, No. 4 (2011), pp. 29-52.
50. Falk, Richard A, 'The Beirut Raid and the International Law of Retaliation', *American Journal of International Law*, Vol. 63, No. 3 (1969), pp. 415-443.

51. Fitzsimmons, Michael, 'The importance of being special: Planning for the future of US special operations forces', *Defense & Security Analysis*, Vol. 19, No. 3 (2003), pp. 203-218.
52. Foley, Charles, *Commando Extraordinary* (London: Cassell, 1998).
53. Gatchel, Theodore L, 'Pseudo Operations: A Double-Edged Sword of Counterinsurgency', in Norwitz, Jeffrey H, *Armed Groups: Studies in National Security, Counterterrorism and Counterinsurgency* (Newport, RI: US Naval War College, 2008), pp. 61-72.
54. Geraghty, Tony, *Who Dares Wins: The Story of the SAS, 1950-1992* (London: Time Warner, 2002).
55. Ghose, R, 'Air Power in Sub-Conventional Warfare in the South Asian Region: Past Lessons and Future Trends', *Air Power*, Vol. 1, No. 5 (2010), pp. 1-30.
56. Gortzak, Yoav, 'Using Indigenous Forces in Counterinsurgency Operations: The French in Algeria, 1954-1962', *Journal of Strategic Studies*, Vol. 32, No. 2 (2009), pp. 307-333.
57. Grau, Lester W and Jalali, Ali Ahmad, 'Forbidden Cross-border Vendetta: Spetsnaz Strike into Pakistan During the Soviet-Afghan War', *Journal of Slavic Military Studies*, Vol. 18, no. 4 (2005), pp. 661-672.
58. Gray, Colin S, 'Handfuls of Heroes on Desperate Ventures: When do Special Operations Succeed?' *Parameters* (Spring 1999), http://strategicstudiesinstitute.army.mil/pubs/parameters/Articles/99spring/gray.htm, 16 September 2013.
59. Griffin, Christopher, 'Major Combat Operations and Counterinsurgency Warfare: Plan Challe in Algeria, 1959-1960', *Security Studies*, Vol. 19, No. 3 (2010), pp. 555-589.
60. Gunaratna, Rohan and Oreg, Aviv, 'Al Qaeda's Organizational Structure and its Evolution', *Studies in Conflict & Terrorism*, Vol. 33, No. 12 (2010), pp. 1043-1078.
61. Gupta, Shekhar, 'Blood, Sweat and Tears', in *The Punjab Story* (New Delhi: Roli, 2004), pp. 73-100.
62. Gupta, Shishir, *Indian Mujahideen: The Enemy Within* (New Delhi: Hachette India, 2011).
63. Haqqani, Husain, *Pakistan: Between Mosque and Military* (Washington DC: Brookings Institution Press, 2005).

64. Harclerode, Peter, *Fighting Dirty: The Inside Story of Covert Operations from Ho Chi Minh to Osama Bin Laden* (London: Cassell, 2001).
65. Hewitt, Steve, *Snitch: A History of the Modern Intelligence Informer* (New York: Continuum, 2010).
66. Hilali, AZ, 'The costs and benefits of the Afghan War for Pakistan', *Contemporary South Asia*, Vol. 11, No. 3 (2002), pp. 291-310.
67. Hoffman, Bruce, *Commando Raids: 1946-1983* (Santa Monica, CA: RAND Corporation, 1985).
68. Hoffman, Bruce; Taw, Jennifer M and Arnold, David, *Lessons for Contemporary Counterinsurgencies: The Rhodesian Experience* (Santa Monica, CA: RAND Corporation, 1991).
69. Hughes, Geraint and Tripodi, Christian, 'Anatomy of a Surrogate: historical precedents and implications for contemporary counter-insurgency and counter-terrorism', *Small Wars & Insurgencies*, Vol. 20, No. 1 (2009), pp. 1-35.
70. Hughes, Geraint, 'The use of undercover military units in counter-terrorist operations: a historical analysis with reference to contemporary anti-terrorism', *Small Wars & Insurgencies*, Vol. 21, No. 4 (2010), pp. 561-590.
71. Ilardi, Gaetano Joe, 'Al Qaeda's Counterintelligence Doctrine: The Pursuit of Operational Certainty and Control', *International Journal of Intelligence and CounterIntelligence*, Vol. 22, No. 2 (2009), pp. 246-274.
72. Ilardi, Gaetano Joe, 'Irish Republican Army Counterintelligence', *International Journal of Intelligence and CounterIntelligence*, Vol. 23, No. 1 (2010), pp. 1-26.
73. Ingram, Martin and Harkin, Greg, *Stakeknife: Britain's Secret Agents in Ireland* (Dublin: O'Brien Press, 2004).
74. Isenberg, David, 'Special Forces: Shock Troops for the New Order', *Middle East Report, No. 177 Arms race or Arms Control in the Middle East?* (1992), pp. 24-27.
75. Johnson, Robert, 'Pakistan's Inter Services Intelligence', in Davies, Philip HJ and Gustafson, Kristian C eds., *Intelligence Elsewhere: Spies and Espionage Outside the Anglosphere* (Washington, DC: Georgetown University Press, 2013), pp. 115-139.

76. Jonas, George, *Vengeance* (London: Harper Perennial, 2006).
77. Jones, Clive, 'A Reach Greater than the Grasp': Israeli Intelligence and the Conflict in South Lebanon 1990–2000', *Intelligence and National Security*, Vol. 16, No. 3 (2001), pp. 1-26.
78. Kapila, Viney, *The Indian Air Force: A Balanced Strategic and Tactical Appreciation* (New Delhi: Ocean Books, 2002).
79. Kenner, Ron (pseudonym), 'Soviet Army Spetsnaz: The Red Elite from the Hysteria of the 1980s to the Present', *Journal for Intelligence, Propaganda and Security Studies*, Vol. 4, No. 1 (2010), pp. 91-105.
80. Khalidi, Ahmed S, 'The War of Attrition', *Journal of Palestine Studies*, Vol. 3, No. 1 (1973), pp. 60-87.
81. Kinnunen, Eero and Grau, Lester W, 'Two Tours in Afghanistan: Twenty Years and Two Armies Apart', *Military Review*, Vol. 91, No. 3 (2011), pp. 40-49.
82. Klein George C, 'Thinking SWAT', *Journal of Police Crisis Negotiations*, Vol. 3, No. 1 (2003), pp. 59-83.
83. Kovacs, Amos, 'Using Intelligence', *Intelligence and National Security*, Vol. 12, No. 4 (1997), pp. 145-164.
84. Kramer, Mark, 'Guerrilla Warfare, Counterinsurgency and Terrorism in the North Caucasus: The Military Dimension of the Russo-Chechen Conflict', *Europe-Asia Studies*, Vol. 57, No. 2 (2005), pp. 209-290.
85. Krause, Michael D, 'Moltke and the Origins of the Operational Level of War', accessed online at http://www.history.army.mil/books/OpArt/germany1.htm, on 29 October 2013.
86. Kumar, A Vinod, 'A Year after 26/11: Soft Responses of a Reluctant State', *Strategic Analysis*, Vol. 33, No. 6 (2009), pp. 800-804.
87. Kutger, Joseph P, 'Irregular Warfare in Transition', *Military Affairs*, Vol. 24, No. 3 (1960), pp. 113-123.
88. LaRaia, William and Walker, Michael C, 'The Siege in Mumbai: A Conventional Terrorist Attack Aided by Modern Technology', in Haberfeld, MR and Hassell, Agostino von eds., *A New Understanding of Terrorism: Case Studies, Trajectories and Lessons Learned* (New York: Springer, 2009), pp. 309-340.
89. Lefevre, Eric, *Brandenburg Division: Commandos of the Reich* (Paris: Histoire & Collections, 2000).

90. Lidder, HS, 'Employment of Special Forces to Achieve Strategic Objectives: Exploiting Opprtunities with a Proactive Approach', in Yadav, VS ed., *Employment of Special Forces: Challenges and Opportunities for the Future* (New Delhi: Vij Publications, 2012), pp. 89-95.
91. Lillerud, Boye, 'Flipping the COIN: Unity of Effort and Special Operations Forces', *NUPI Working Paper No. 752* (2008).
92. Livingstone, Neil C, 'Proactive Responses to terrorism: Reprisals, Preemption and Retribution', in Livingstone, Neil C and Arnold, Terell E, *Fighting Back* (Lexington, MA: Lexington Books, 1986), pp. 109-131.
93. McKinney, Mike and Ryan, Mike, *Chariots of the Damned: Airborne Rescues from Vietnam to Kosovo* (London: HarperCollins, 2001).
94. McNab, Andy, *Immediate Action* (London: Corgi, 1995).
95. McRaven, William H, *Spec Ops: Case Studies in Special Operations Warfare—Theory and Practice* (Novato, CA: Presidio, 1996).
96. Mir, Amir, *Talibanization of Pakistan: From 9/11 to 26/11* (New Delhi: Pentagon Security International, 2009).
97. Moore, David M; Allen, David and Antill, Peter D, 'Strategy development for special operations force logistics', *Defence Studies*, Vol. 3, No. 1 (2003), pp. 66-108.
98. Myers, Joseph C, 'Review Essay: The Quranic Concept of War', *Parameters* (Winter 2006-07), pp. 108-127.
99. Nandy, Bibhuti Bhusan, 'DGFI: State Within a State', in Saikia, Jaideep ed., *Bangladesh: Treading the Taliban Trail* (New Delhi: Vision Books, 2006), pp. 173-187.
100. Nayyar, AH and Salim, Ahmed, *The Subtle Subversion: The State of Curricula and Textbooks in Pakistan* (Islamabad: Sustainable Development Policy Institute, 2006).
101. O'Ballance, Edgar, *Afghan Wars: Battles in a Hostile Land, 1839 to the Present* (London: Brassey's, 2002).
102. Orwin, Etan M, 'Squad leaders today, village leaders tomorrow: Muslim auxiliaries and tactical politics in Algeria, 1956-1962', *Small Wars & Insurgencies*, Vol. 23, No. 2 (2012), pp. 330-351.
103. Owen, Mark with Maurer, Kevin, *No Easy Day* (London: Penguin, 2012).

104. Pahlavi, Pierre, 'Political Warfare is a Double-edged Sword: The Rise and Fall of the French Counterinsurgency in Algeria', *Canadian Military Journal* (2007-2008), pp. 53-63.
105. Pathak, DC, 'Understanding the "War on Terror"', *Eternal India* (2009), pp. 46-54.
106. Plaster, John L, *SOG: The Secret Wars of America's Commandos in Vietnam* (London: Onyx, 1998).
107. Plater-Zyberk, Henry, *Russia's Special Forces*, Conflict Studies Research Centre, September 2005, http://studies.agentura.ru/centres/csrc/rsp.pdf, 24 October 2013.
108. Pedahzur, Ami, *The Israeli Secret Services and the Struggle against Terrorism* (New York: Columbia University Press, 2010).
109. Petrelli, Niccolo, 'The missing dimension: IDF special operations forces and strategy in the Second Lebanon War', *Small Wars & Insurgencies*, Vol. 23, No. 1 (2012), pp. 56-73.
110. Porch, Douglas, 'The dangerous myths and dubious promise of COIN', *Small Wars & Insurgencies*, Vol. 22, No. 2 (2011), pp. 239-257.
111. Rajagopalan, Rajesh, *Fighting Like a Guerrilla: The Indian Army and Counterinsurgency* (New Delhi: Routledge, 2008).
112. Rassler, Don; Fair, C Christine; Ghosh, Anirban; Jamal, Arif and Shoeb, Nadia, *The Fighters of Lashkar-e-Taiba: Recruitment, Training, Deployment and Death*, Combating Terrorism Center, http://www.isn.ethz.ch/Digital-Library/Publications/Detail/?id=163497, 8 November 2013.
113. Raza, Maroof, 'Introduction', in Raza, Maroof ed., *Confronting Terrorism* (New Delhi: Penguin, 2009), pp. xix-xxxv.
114. Reid-Daly RF, *PamweChete: The Legend of the Selous Scouts* (Welteverdenpark: Covos-Day, 2001).
115. Rennie, James, *The Operators: The Army's Top Secret Elite* (London: Century, 1996).
116. Scahill, Jeremy, *Dirty Wars: The World is a Battlefield* (London: Serpent's Tail, 2013).
117. Schofield, Carey, *Inside the Pakistan Army* (New Delhi: Pentagon Security International, 2011).

118. Schultz, Richard H, Jr, *The Secret War Against Hanoi: Kennedy's and Johnson's Use of Spies, Saboteurs, and Covert Warriors in North Vietnam* (New York: HarperCollins, 1999).
119. Shahzad, Saleem, *Inside Al Qaeda and the Taliban* (London: Pluto Press, 2011).
120. Sharma, LN and Kumar, Umesh, 'A Need for a Cogent Counter-Terror Doctrine', *Eternal India* (2009), pp. 81-96.
121. Shlaim, Avi and Tanter, Raymond, 'Decision Process, Choice, and Consequences: Israel's Deep-Penetration Bombing in Egypt, 1970', *World Politics*, Vol. 30, No. 4 (1978), pp. 483-516.
122. Shpiro, Shlomo, 'Israeli Intelligence and Al Qaeda', *International Journal of Intelligence and CounterIntelligence*, Vol. 25, No. 2 (2012), pp. 240-259.
123. Skelly, Joseph Morrison, 'A Review of The Activity's Active Intelligence', *International Journal of Intelligence and CounterIntelligence*, Vol. 21, No. 2 (2008), pp. 418-427.
124. Smith, Michael, *Killer Elite: The Inside Story of America's Most Secret Special Operations Team* (Weidenfeld & Nicolson).
125. Smith, Steve, 'Groupthink and the Hostage Rescue Mission', *British Journal of Political Science*, Vol. 15, No. 1 (1985), pp. 117-123.
126. Sood, VK and Sawhney, Pravin, *Operation Parakram: The War Unfinished* (New Delhi: Sage, 2003).
127. Shekhar, Chandra, 'Special Forces in India: Doctrinal Issues and Employment Concepts', in Vijay Oberoi ed., *Special Forces: Doctrine, Structures and Employment Across the Spectrum of Conflict in the Indian Context* (New Delhi: Knowledge World, 2006), pp. 147-15.
128. Stanton, Shelby L, *Green Berets at War: US Army Special Forces in Southeast Asia, 1956-1975* (New York: Ivy Books, 1985).
129. Stringer, Kevin, 'Historical Lessons from the British and Germans: The case for integrated indigenous forces as a logical force multiplier', *Canadian Military Journal* (2007), pp. 55-62.
130. Strobl, Staci and Lindsay, Jon R, 'Lost in Transition: Khobar Towers and the Ambiguities of Terrorism in the 1990s', in Haberfeld, MR and Hassell, Agostino von eds., *A New Understanding of Terrorism: Case Studies, Trajectories and Lessons Learned* (New York: Springer, 2009), pp. 283-307.

131. Subramaniam, Arjun, 'Air Power in Stability and Anti-Terrorist Operations', *Air Power*, Vol. 6, No. 3 (2011), pp. 25-51.
132. Subramaniam, Arjun, 'Challenges of Protecting India from Terrorism', *Terrorism and Political Violence*, Vol. 24, No. 3 (2012), pp. 396-414.
133. Sutton, Keith, 'Army Administration Tensions over Algeria's Centres de Regroupment, 1954-1962', *British Journal of Middle Eastern Studies*, Vol. 26, No. 2 (1999), pp. 243-270.
134. Swami, Praveen, *India, Pakistan and the Secret Jihad: The Covert War in Kashmir 1947-2004* (New York: Routledge, 2007).
135. Terrill, W Andrew, 'The nature and value of commando operations during the Egyptian-Israel war of attrition', *Small Wars & Insurgencies*, Vol. 8, No. 2 (1997), pp. 16-34.
136. Thomas, David, 'The Importance of Commando Operations in Modern Warfare 1939-82', *Journal of Contemporary History*, Vol. 18, No. 4 (1983), pp. 689-717.
137. Thomas, Gordon, *Gideon's Spies: Mossad's Secret Warriors* (London: Pan, 1999).
138. Urban, Mark, *Task Force Black: The Explosive True Story of the SAS and the Secret War in Iraq* (London: Abacus, 2011).
139. Weale, Adrian, *Secret Warfare: Special Operations Forces from the Great Game to the SAS* (London: Coronet, 1997).
140. Waller, Douglas C, *The Commandos: The Inside Story of America's Secret Soldiers* (New York: Dell, 1995).
141. Wege, Carl Anthony, 'Hizballah's Counterintelligence Apparatus', *International Journal of Intelligence and CounterIntelligence*, Vol. 25, No. 4 (2012), pp. 771-785.
142. Weingartner, James J, 'Otto Skorzeny and the Laws of War', *Journal of Military History*, Vol. 55, No. 2 (1991), pp. 207-224.
143. Witty, David M, 'A Regular Army in Counterinsurgency Operations: Egypt in North Yemen, 1962-1967', *Journal of Military History*, Vol. 65, No. 2 (2001), pp. 401-439.
144. Yousaf, Mohammad and Adkin, Mark, *Afghanistan the Bear Trap: The Defeat of a Superpower* (Barnsley, South Yorkshire: Leo Cooper, 1992).
145. Zaidi, S Hussain, *Black Friday: The True Story of the Bombay Bomb Blasts* (New Delhi: Penguin, 2002).

146. Zaidi, S Hussain with Bhatt, Rahul, *Headley and I* (New Delhi: HarperCollins, 2012).

Official Documents

Review of Terrorist Activities in Punjab for the month of September 1991, compiled by the Office of the Additional Director-General of Police Intelligence.

Review of Terrorist Activities in Punjab for the year 1997, compiled by the Office of the Additional Director-General of Police Intelligence.

Websites

Air Force Magazine
Bharat-Rakshak
Der Spiegel
Indian Express
New York Times
Observer Research Foundation
South Asia Analysis Group
South Asia Terrorism Portal
The Atlantic Monthly
Times of India
Times of Israel
YouTube

Index

A

Abbottabad xviii, 53, 54, 64
Abwehr 74
AC-130 Gunship 64
Active Service Units (ASUs) 32
Aerospace Rescue and Recovery Service 52
Aerospace Rescue and Recovery Squadron (ARRS) 52
Afghan Forces 22
Afghanistan vii, 6, 32, 46, 47, 49, 50, 53, 97, 107, 109, 110, 111, 112, 113, 116, 119, 121, 122, 123, 124, 157
Ahmed, Waqar 97
Aideed, Mohammad Farrah xix, 48
Airborne Warning and Control Systems (AWACS) 64
Air France xxiv
AK-74 Rifle 122
AKM Rifles 122
Algeria xxii, 8, 9, 13, 14, 15, 16
Algerian Independence 8
Algerian War 14, 168
Ali, Qurban 88
Allied Powers 43
Al Qaeda xxii, xxiii, 31, 32, 80, 95, 105, 107, 136, 145
Al-Wazir, Khalid 161
American xvi, xx, 8, 11, 13, 19, 21, 22, 23, 25, 26, 30, 32, 33, 36, 40, 41, 45, 46, 48, 50, 54, 55, 62, 63, 64, 69, 74, 83, 95, 122, 132, 133, 138, 141, 149, 150, 155, 156, 163, 164, 165, 168
Anglo-American model 46, 69, 83, 164, 168
Anglo-Saxon model 45
Anti-Terrorist Squad (ATS) 152
Arab-Israeli conflict xii, 127, 141, 161, 168
Arafat, Yasser 137
ARCLIGHT missions 47
Area Denial 7, 8, 16, 21
Area Dominance 4, 7, 13, 16, 112, 113, 114, 141, 142, 150, 151, 154, 166
Area Penetration 4, 7, 17, 20, 21, 22, 112, 121, 127, 131, 141, 142, 150, 151, 156, 163, 165, 166, 167
Argentina 46, 138
Army Rangers 39
Arreguin-Toft, Ivan 13
Assessment and Selection (A&S) 79
Australia xvii, xxvii, 167

B

B-52 bombers 47
Ba'ath Party 13
Bahawalnagar 159

Bahawalpur 159
Bakshi, GD 155
Bangladesh 106, 162
Bau-Lehr Bataillon 800 74
Beirut 27, 31, 138, 143, 156
Black Hawk Down xix, 49
Black September 138, 139, 140, 141, 143
Blackwater 22
Blitzkrieg doctrine 5, 56
Boer War xx, xxi
Brandenburg 56, 74, 75
British xiii, xx, xxi, xxiv, xxv, xxviii, xxix, 4, 5, 6, 8, 32, 35, 42, 45, 46, 51, 58, 67, 68, 69, 71, 74, 75, 76, 78, 89, 90, 125, 165
British Army 4
British SAS 71, 165
Buenos Aires 46
Bush War 61, 63, 168

C

Cairo 130, 131, 135
Cambodia 18, 19, 20, 47, 62
Canada xxvii
Central Intelligence Agency (CIA) 17, 18, 19, 31, 39, 40, 41, 96, 107, 117, 119, 120, 123, 141, 160, 163
CH-47 Chinooks 64
Challe, Maurice (Gen) 14
Challe Plan 13, 14, 15, 16
Chechen 30, 49
Chechnya 49, 50
China x, xxviii, 18, 63, 128
Churchill, Winston 5
Civilian Irregular Defense Group (CIDG) 10, 11
Clausewitzian model 90
Close-quarter Battle (CQB) 80
Cold Start Doctrine 162
Cold War 37, 45, 125, 157
Combating Terrorism Center 159
Command and Control (C2) xv, 54, 64, 108, 114, 164
Commando Raids 1946-1983 xxix
Commandos de Chasse 15

D

Dawson's Field crisis 28
Delta Force 22, 39, 42, 71, 77, 79, 80, 165
Deobandi 106
Douglas, McDonnell 530 54
Durand Line 110

E

EC-130E/H electronic warfare aircraft 64
Egypt 12, 128, 129, 130, 131, 132, 133, 134, 135, 154
Egyptian Army 12
Entebbe xxiv, 144
Escape and evasion (E&E) 70
Escobar, Pablo 42
Establishment 22 165, 166
Europe 4, 5, 30, 31, 57, 88, 140, 141

F

F-4 Phantom fighters 135
F-15C fighters 64
Faisalabad 159
Federal Bureau of Investigation (FBI) 33
Federally Administered Tribal Areas (FATA) 64
Federal Security Service (FSB) 29, 30
Fidayeen xvi, 55, 56, 80, 101, 107, 108, 149, 152, 153, 154, 160
First Intifada 141
Force Research Unit (FRU) 35, 76, 77, 78, 143
France xxiv, 8, 9, 14, 15, 78
Franco-Algerian War 168
French Army 8, 9, 13, 14, 15

G

German Panzer formations 56
German Wehrmacht 4, 5, 45, 69, 74
Ghadir 37
Ghatak platoons 152

Gilani, Daood/Headley, David 45, 95, 152, 161
Gorbachev, Mikhail 112
Green Berets 1, 9, 11, 21, 37, 47
GRU (Soviet Military Intelligence) 116
Gujranwala 159

H
Hadith, Ahle 105, 106
Harkat-ul Jihad Islami-Bangladesh (HuJI-B) 106
Harvard 13, 115
High-Altitude/Low-Opening (HALO) 71
Hitler, Adolf 74
Hizballah xxii, 28, 31, 32, 33, 34, 46, 128, 129, 137, 138, 142, 143, 160
Ho Chi Minh 18, 19, 20, 36, 47, 120
Hoffman, Bruce xxv, xxix, 143
Huey helicopters 41
Hussein, Qusay 42
Hussein, Saddam 42
Hussein, Uday 42

I
Ibrahim, Dawood 98
Improvised Explosive Devices (IEDs) 101, 123
India iv, xii, xiii, xxvi, xxvii, xxviii, 6, 17, 21, 22, 23, 28, 42, 43, 46, 55, 56, 58, 62, 63, 64, 65, 67, 69, 78, 80, 84, 87, 88, 89, 90, 91, 92, 94, 95, 96, 97, 98, 99, 100, 101, 102, 103, 104, 105, 106, 107, 108, 109, 110, 111, 123, 124, 125, 126, 128, 129, 133, 134, 135, 136, 137, 139, 140, 145, 147, 149, 150, 151, 152, 153, 154, 155, 156, 157, 158, 160, 161, 162, 163, 164, 165, 166, 167, 168
Indian Air Force (IAF) 48, 64, 65
Indian Army ix, xxvi, xxviii, xxx, 64, 83, 124, 136, 165, 166
Indian Mujahideen (IM) 87, 94, 95, 96, 103, 105, 106, 154, 165
Instrumentalist School 88
Intelligence Support Activity (ISA) 39, 40, 41, 42, 78, 79, 165
Inter Services Intelligence (ISI) 46, 55, 87, 88, 89, 90, 91, 94, 95, 96, 97, 98, 99, 100, 101, 103, 104, 105, 106, 107, 108, 109, 110, 111, 112, 117, 119, 120, 122, 123, 124, 125, 137, 138, 139, 145, 149, 150, 155, 160, 161, 162, 164
Irish Republic Army (IRA) 32, 35, 76, 77
Irredentist School 88, 89
Irregular Warfare xii, xiv, 34
Islam 88, 100, 106
Islamabad 40, 90, 105, 109, 110, 111, 125, 140, 156, 157, 158
Islamic Jihad Organization (IJO) 138
Israel xxiii, xxiv, 11, 28, 73, 81, 82, 100, 126, 128, 129, 130, 131, 132, 133, 134, 135, 136, 137, 140, 141, 142, 143, 144, 145, 147, 150, 154
Israeli Mossad xxiii
Israel-Palestine conflict 126

J
Jaffna University 65
Jalalabad Offensive 1988 122
Jamaat-e-Islami (JeI) 97
Jammu and Kashmir (J&K) 88, 89, 92, 98, 102, 110, 123, 124, 125, 128, 129, 139, 151, 157, 158, 159, 162
Jihad/Jihadism/Jihadist/Jihadist Group x, xiii, xxvii, 13, 80, 87, 88, 95, 96, 100, 101, 104, 105, 106, 107, 108, 110, 123, 125, 139, 149, 150, 155, 158, 161
Joint Special Operations Command (JSOC) 21, 22, 34, 51
Jordan 144
Jundal, Abu 145

K
Karachi xii, 56, 136, 159
Kargil ix, xxx, 55
Karim, Abdul (Tunda) 145

Kashmir 97, 98, 99, 110, 139
Kasab, Ajmal 105
Kasur 159
Katiba 15
Kenya 144
Khalistan, Kashmir and Muslim Militancy (K2M) 98
Krer Raid 117

L
Laden, Osama bin xviii, xxiii, 4, 22, 27, 104, 144, 157
Lahore xii, 159
Landing Zone (LZ) 20, 47, 48
Laos 18, 19, 20, 47, 62
Lashkar-e-Taiba (LeT) 45, 46, 55, 56, 63, 80, 87, 94, 95, 96, 100, 101, 103, 105, 106, 107, 108, 129, 138, 139, 145, 152, 153, 154, 156, 157, 158, 159, 160, 161, 163, 164
Latin America 13, 33, 46, 78
Lebanese Civil War 26
Lebanese Islamists 137
Lebanon 26, 27, 28, 32, 128, 137, 142, 143, 161
Line of Control (LoC) 63, 128, 133, 135, 136, 145, 166
Little Bird 54

M
Malaya xxi, xxv, 167
Malayan Emergency xxix
Maldives 107
Man-portable Air Defence Systems (MANPADS) 46, 48, 49, 110
Marine Commandos (MARCOS)/Marines 11, 17, 26, 27, 56, 138
Masoud, Ahmed Shah 114
MC-130 command and control aircraft 64
Mehsud, Baitullah 160
MH-60 helicopters 53, 64
MI-24D Hind gunships 116
Middle East 6, 27, 40, 46, 78, 105, 107, 141, 144, 161
Ministry of Defence (MoD) 64
Mizo 97
Mogadishu xix, 48
Mogadishu, battle of 48
Mohammad, Fasih 145
Moltke, Helmuth von 150
Money, Ideology, Compromise and Ego (MICE) 37
Montagnard 9, 10, 11
Morocco 14
Moscow 112, 121, 123, 135
Mossad xxiii, xxiv, 31, 46, 140, 141, 144, 145, 160, 161
Mozambican Army 59, 60, 61
Mozambique 39, 59, 60
Mughniyeh, Imad 138, 160, 161
Mujahideen 112, 113, 114, 115, 116, 117, 118, 119, 120, 121, 122, 123
Multan 159
Mumbai xii, xxvi, 55, 56, 95, 98, 99, 104, 105, 136, 151, 152, 153, 154, 155, 157, 161
Mussolini, Benito 43, 57

N
Naga 97
Nap-of-the-earth (NOE) 53, 63
Nariman House 161
National Security Advisor (NSA) 166
National Security Agency 41
National Security Council 50
National Security Guard (NSG) 81, 82, 152, 153
Nasser, Gamal Abdel (President) 132
Naval Special Warfare Development Group (DEVGRU—SEAL Team 6) 165
Nazi 4
Nepal 106
New Delhi iv, xii, xxvi, xxvii, xxviii, 6, 48, 64, 78, 80, 82, 92, 96, 97, 98, 101, 102, 103, 104, 105, 106, 108, 109, 117, 123, 151, 166, 167
New York 1, 19, 21, 30, 32, 33, 34, 53, 68, 79, 82, 89, 107, 143, 144, 151, 163

New Zealand xxv, xxvii
Nile 130
Northern Ireland 32, 35, 75, 76, 77, 78, 168
North Vietnam 17, 18, 19, 47, 51, 52, 55, 62
North Yemen 12, 13
Norway 140

O
Obama, Barack 22, 157
Omsk vans 116
Operational Plan (OPLAN) 34A 18
Operations
  Operation Barrier 115
  Operation Bayonet 140, 141, 143
  Operation Brasstacks (1987) 162
  Operation Curtain 115
  Operation Eagle Claw 39, 50
  Operation Eland 58
  Operation Irene 48
  Operation Parakram 64, 163
  Operation Shock 130
  Operation Spring of Youth 143
  Operation Thunderbolt xxiv
Operations Wing/OP-37 55

P
Pacific Ocean 52
Pakistan vii, ix, x, xii, xiii, xxvi, xxvii, xxviii, xxix, xxx, 7, 21, 22, 40, 43, 46, 56, 58, 63, 64, 65, 67, 82, 83, 87, 88, 89, 90, 91, 92, 94, 95, 97, 98, 100, 101, 102, 104, 105, 106, 107, 108, 109, 110, 111, 112, 115, 116, 117, 118, 119, 120, 121, 122, 123, 124, 125, 126, 128, 129, 133, 134, 135, 136, 137, 138, 139, 140, 145, 147, 148, 149, 150, 152, 154, 155, 156, 157, 158, 159, 161, 162, 163, 164, 165, 166, 167, 168
Pakistan Army xiii, 90, 91, 94, 107, 109, 111, 124, 149, 159, 163
Pakistan Intelligence Bureau (IB) 88, 89
Pakistani Special Services Group 80
Pakistan Military Academy 104
Pakistan Occupied Kashmir (PoK) 136, 137, 158, 159, 166, 167
Palestinian Authority (PA) 137
Palestinian Liberation Organization (PLO) 128, 129, 137, 138, 139, 141, 143
Pan-Islamist militants 90
Panjshir Valley 114
Parachute Regiment 71
Partition 88
Pashtuns 110
Pentagon 17, 18, 22, 109, 117
Persian Gulf 36, 37, 54, 90, 100
PNS Iqbal 56
Popular Defence Force 131
Popular Front of India (PFI) 106
Prisoner-of-War (POW) 51
Punjab 6, 82, 97, 100, 139, 151, 159, 162

Q
Quadrillage 14, 15
Quran 90
Quranic Concept of War, The 90

R
Rana, Tawahhur 95
Rangers 71
Rashtriya Rifles (RR) 124, 125
Rawalpindi 110
Reagan, Ronald (President) 26
Reich 56
Research & Analysis Wing (RAW) 164
Reserves Generales (RG) 14, 15
Reversible Strengths xv
Rhodesia xxv, 38, 59, 60, 61, 62, 167
Rhodesian Army 58, 61, 71
Rhodesian Bush War 168
Rumsfeld, Donald 22
Russian Spetsnaz 72

S
Saigon 148, 149
Salafist school 88, 89

Salameh, Ali Hassan 138, 140, 141, 161
Salang Tunnel 122
Salvador Model 13
Saudi Arabia 12, 90
Sayaret Matkal 72, 81, 82, 131
SEAL Team 6 22, 23, 65, 165
Selous Scouts 38, 39, 58, 59, 60, 61, 71, 72, 79
Shayetet 13 131, 142
Sinai 128
Sindh 162
Six Day War of 1967 128
Skorzeny, Otto 57, 58
Somalia xix, xx
Somali pirates 23
Son Tay raid 1970 51
South Africa xx
South America 46
South Asia 91, 96, 101, 105, 106, 110, 138, 150
Southeast Asia 1, 78
South Vietnam 9, 17, 18, 55, 149
Soviet/Soviets xii, 28, 46, 49, 73, 89, 97, 107, 108, 109, 110, 111, 112, 113, 114, 115, 116, 117, 118, 119, 120, 121, 122, 123, 124, 127, 128, 131, 132, 133, 135, 137, 157, 168
Soviet-Afghan War xii, 89, 107, 108, 109, 110, 119, 127, 168
Soviet Spetsnaz/Spetsnaz 28, 49, 50, 72, 73, 109, 112, 113, 114, 115, 116, 117, 118, 119, 120, 121, 122
Soviet Union 49, 109, 112, 120, 122, 124, 128, 132, 133, 135, 137
Special Administrative Sections 8
Special Air Service (SAS) 4, 5, 6, 8, 9, 15, 34, 70, 71, 76, 77, 78, 79, 80, 117, 165
Special Heli-borne Operations (SHBOs) 65
Special Operations Command (SOCOM) 21, 34, 51, 64
Special Operations Forces/Special Forces (SOF) vii, ix, xiv, xvii, xviii, xix, xx, xxiv, xxv, xxix, 1, 4, 6, 8, 9, 10, 13, 21, 37, 39, 40, 42, 46, 47, 48, 49, 52, 67, 68, 69, 70, 71, 72, 73, 74, 76, 78, 79, 80, 81, 82, 83, 84, 121, 127, 129, 131, 132, 135, 142, 143, 148, 150, 163, 164, 165, 166, 167
  9 PARA Special Forces 166
  14th Intelligence Company (14 Int) 35, 76, 77, 76, 78, 143
  18th United Kingdom Special Forces Signal Regiment 42
  51st and 52nd Special Action Groups 82
  75th Ranger Regiment 22
  160th Special Operations Aviation Regiment (160th SOAR) 51
Special Operations Group 151
Special Ranger Groups 82
Special Reconnaissance Regiment 42
Special Research Apparatus (SRA) 138
Special Services Group (SSG) 80, 117, 159, 160, 161, 164
Special Weapons and Tactics (SWAT) xvi, 108, 152, 153, 166
Spring of Youth 143
Sri Lanka 107
Students' Islamic Movement of India (SIMI) 98, 106
Studies and Observations Group—Military Assistance Command Vietnam (SOG-MACV) 17, 18, 19, 20, 36, 47, 71
Sundarji Doctrine 162
SVD sniper rifles 122

T

Tamil Tigers 65
Tasking, Intelligence, Logistics and Training (TILT) xii
Teheran 39
Terror attacks
  9/11 (September 11) attacks 96, 124
  26/11 Mumbai terror attacks xxvi, xxvii, 80, 95, 102, 106, 107, 117, 125, 134, 136, 139, 145, 151, 152, 153, 154, 156, 157, 159, 161, 163

Munich Massacre, Olympics 1972 137, 140, 161
Tel Aviv Savoy Hotel attack, 1975 161
Tet Offensive 148, 149
Thirty Years War 88
Trail Watchers 20
Trojan Horse 56
Tunisia 14, 161

U
Ugandan 144
United Kingdom (UK) xxvii, 6, 69, 125, 154, 155
United Nations General Assembly 137
United States (US) xviii, xix, xxvi, xxvii, 1, 4, 6, 9, 10, 11, 13, 17, 18, 19, 20, 21, 22, 23, 26, 27, 31, 33, 34, 36, 37, 38, 39, 40, 41, 42, 47, 48, 49, 50, 51, 52, 54, 55, 62, 63, 64, 67, 68, 69, 71, 78, 79, 80, 85, 93, 95, 96, 98, 99, 100, 117, 124, 125, 133, 135, 136, 137, 141, 148, 149, 150, 154, 155, 156, 157, 159, 160, 165
Unmanned Aerial Vehicles (UAVs) 83
US Air Force 47, 48
US Army Special Forces 1, 9, 21, 37, 40, 52, 71
US embassy 27, 39, 148, 149
US Navy SEALs 36, 68, 71, 80
USS Carl Vinson 64
USS Enterprise 64

V
Vietnam 9, 17, 18, 19, 20, 48, 49, 55, 62, 149
Vietnam War 9, 11, 17, 19, 36, 47, 148, 168

W
Wacht am Rhein offensive 57
Wahhabi 105
War of Attrition 128, 130, 133, 134, 136
War Paradigm 162
Washington 22, 36, 90, 123, 133, 157
Weinberger, Caspar 26
Western Europe 57, 140
White House 50
Wile E Coyote-style attacks 107
Winter Harvest 41
World Trade Center 22
World War xxviii, 47
World War II (WWII) 4, 45, 46, 67, 75, 78, 83

Y
Yemeni Campaign 12

Z
Zambia 39
Zimbabwe xxv, 38